The Biopsychosocial-Spiritual

Approach:

Towards a Holistic Understanding and

Treatment of Drug Addiction

i

The Biopsychosocial-Spiritual Approach:

Towards a Holistic Understanding and Treatment of Drug Addiction

Rev. Dr. Joyzy Pius Egunjobi

Printed in the United States of America by Lulu Press Inc.

ISBN: 978-1-365-12160-9

Joyzy Pius Publications

www.Lulu.com/joyzypbooks
joyzyp@yahoo.com
joyzyp71@gmail.com

Publications

DEDICATION

This work is dedicated to those who struggle day and night with the demon called Addiction in all its various forms.

TABLE OF CONTENTS

ACKNOWLEDGEMENTS

"I will praise you, Lord my God, with all my heart; I will glorify your name forever. For great is your love toward me…" (Ps 86:12-13) – and for the blessings of your wisdom, I will exalt your Holy Name.

I will continue to be grateful to my late father, Chief Emmanuel Olayiwola Egunjobi and my sweet mother, Chief Mrs Mary Faderera Egunjobi for providing such a functional family environment and placing my feet on the path of righteousness and education. Same thanks go to my siblings, Joseph Segun Egunjobi, Catherine Funmilayo Oriajoye and Felix Babatunde Egunjobi, and their families for their constant prayers and encouragements. I love you all.

To my Bishop, Most Rev Dr Emmanuel Adetoyese Badejo, for his fatherly support, opportunity, and encouragements – I say thank you. Also to my pastors, Rev Frs. John Robert Blaker and John Direen for their accommodation, love, and understanding during the course of my Doctoral Studies and Post-Doctoral Clinical Experience. Thank you.

My appreciation also goes to Dr. Michael J. Breining, the Graduate Degree Program Coordinator, Breining Institute, Orangevale, CA, USA who guided me through my Doctoral Program and moderated my Dissertation. I am very grateful for all your kind and encouraging words.

I also appreciate the prayerful and kind supports of Dr. Franca Niameh, who, despite her very tight schedule, took time to write forewords to this work. May the Lord continue to find time for you and nourish you with his love and mercy.

My gratitude also goes to Tom Gorham, the Executive Director of Options Recovery Services, and all the Staff for their supports and understanding in the course of my Post-Doctoral Clinical.

I am also grateful to Oluwafeyikemi 'Bunmi Adeboye who spent time during her vacation to the United States from Portugal proofread the manuscripts and offered valuable corrections and contributions.

May the Lord bless and sanctify all my well-wishers.

Rev Dr. Joyzy Pius Egunjobi

Forward

This informative and engaging understanding of addiction from a bio-psychosocial-spiritual model is an important attempt to understand the complexity of addiction. Several studies have endorsed bio-psychosocial model as a comprehensive framework in understanding addiction (e.g. Griffiths, 2005; Hatala, 2012; Garlland, E.L., et al, 2010). Addiction arises once one has lost the choice or volition, drinks, smokes, or uses drug regardless of the adverse impact. Providing a bio-psychosocial approach is consistent with Griffiths (2005) suggestions whose model of addiction encompasses viewing addiction from six components which are:

> - *Salience*: referring to being preoccupied with a particular activity that one's thoughts and behavior centers excessively on that activity that become more salient and prevent the individual from engaging in other activities.
> - *Mood modification:* refer to the addict engaging in whatever is the drug of choice with the aim to distress, relax or escape from emotional or psychological pain.

- ➢ *Tolerance:* involves a process whereby the amount of whatever is the drug of choice is increased to achieve the initial effects.
- ➢ *Withdrawal symptoms:* refer to the discomfort and unpleasant feeling experienced when the particular drug of choice is discontinued or reduced.
- ➢ *Conflict:* refers to disconnection from self within the abuser and his/her disconnection in interpersonal relationships.
- ➢ *Relapse:* "refer to the tendency for repeated reversion to earlier patterns" (Griffiths, 2005, p. 195) despite abstinence. It should be noted also that all of Griffiths components (salience, mood modification, tolerance, withdrawal symptoms, conflict, and relapse), indicate that an addictive individual continues using the substances in spite of significant substance related problem.

In addition, the *Diagnostic and Statistical Manual of Mental Disorder-Fifth Edition (DMS-5)* recognizes ten separate categories of drugs namely: alcohol, caffeine, hallucinogens, inhalants, opioids, sedatives, hypnotics/anxiolytics, stimulants, cocaine, tobacco, and other unknown substances which are classified under substance-related disorders. According to DSM-5,

> All drugs that are taken in excess have in common direct activation of the brain reward system, which is involved in the reinforcement of behaviors and the production of memories. They produce such an intense activation of the reward system that normal activities may be neglected.... The pharmacological mechanisms by which each class of drugs produces reward are different, but the drug typically activates the system and produce feelings of pleasure, often referred to as a "high". Furthermore, individuals with lower levels of self-control, which may reflect impairments of brain inhibitory mechanisms, may be particularly predisposed to develop substance use disorders, suggesting that the roots of substance use disorders for some persons can be seen in behaviors long before the onset of actual substance use itself. (DSM-5, p. 481)

The biological basis of addiction as described in DSM-5, clearly points out that addiction is not only maintained by the brain reward pathway, but that some individuals have vulnerability that is biological in nature towards addictions.

Rev. Dr. Joyzy Pius Egunjobi's position also addresses biological approach which does support findings that reference the brain reward system or what is called the dopamine hypothesis particularly with rat studies that drug abuse activates brain reward pathway. For example, studies has shown that "mice lacking the dopamine D2 receptor consume less alcohol than normal mice" (Robbins, T.W. & Everitt, B.J, 1999); suggesting that treatment with drug that stimulates dopamine D2 receptor may reduce addictive behavior or contribute to abstinence.

Alongside biological vulnerability are psychosocial or environmental factors that interplay with biology to influence addiction. Psychosocial factors as rightly pointed out includes the individuals' personal experiences, family upbringing, race, ethnicity, interpersonal relationships emotional destress, anxiety, depression, or negative affect.

> Several treatment studies have shown that higher levels of psychological withdrawal or "abstinence" symptoms, such as subjective distress, irritability, drug cravings, sleep, and cognitive problems, occurring during early drug abstinence, even beyond the acute withdrawal phase, are associate with worse treatment outcomes among smokers, cocaine addicts, heroin-dependent individuals, and alcoholics (Sinha, 2011, p.3)

Stress, negative affect, and depression contribute to alcohol or drug cravings, as well as drug relapse. Therefore, biological stress response, psychological and impoverished social environment are indeed predictive of addiction and relapse.

More so, given that addiction is inherently rewarding is not only informative, but makes clear Miller's (1998) argument that "addiction is fundamentally a motivational problem.... The term 'motivation' here is understood not simplistically as will power, but as involving complex biopsychosocial factors" (Millier, 1998, p. 4). The notion here is that addiction is a goal directed behavior that with time and persistence, becomes habitual and reinforces like food (Cardinla, R.N. & Everitt, B.J., 2004). Since addiction is inherently reinforcing, Miller (1998) suggested a motivational model of addiction that utilizes a motivational interviewing as a competing alternative to addiction. Biopsychosocial-spiritual treatment is a motivational approach that motivates the addicts to step inside him/herself to find meaning, value, and self-efficacy. This is where spirituality comes in as a necessary competing motivating construct to be added to the bio-psychosocial model in motivating change and self-efficacy.

The argument for a spiritual approach offers alternatives to addiction that is more motivating than addiction itself. Spirituality touches on the highest level of the human person, a subjective experience (Galanter, 2006) that when embraced, drives the individual biologically, cognitively, behaviorally, psychologically, and socially, toward change. For example, "a unique relationship has been observed between 5-HT receptor density in the forebrain, as recorded by position emission tomography, and the trait of spiritual acceptance, as measured by a standardized personality inventory" (Galanter, 2006, p. 287) indicating that spiritual experience also impact the brain towards positive change that may compete with the reward pathway of addiction. Although Galanter (2006) also pointed out that "research drawn from a variety of different sources suggests that the subjective experiences associated with spiritual or religious phenomena (found in the brain) can potentially be understood in terms of related biological mechanisms" (p. 287). This points to me that spirituality is an effective and powerful route that creates a protective change for confronting addiction. Spiritual approach has been shown to work with the adaptation of the AA twelve-step approach. Also, in my clinical work with patients in an integrative medical setting, patients grounded in spiritual meaning tend to be more motivated towards healing and recovery. Moreover, a study done on *Spiritual direction in addiction treatment: Two Clinical Trials* revealed "...improved outcomes with spiritual direction was based on the widely-expressed view that spirituality is a key component of recovery from addiction. Particularly within the program of AA, continued practice of spiritual disciplines such as prayer and meditation is recommended and regarded as important to maintain conscious contact with God as well as sobriety (Miller, W.R., Forcehimes, A., O'Leary, M., & LaNoue, M.D., 2008).

Therefore, awareness that spirituality is important in approaching addiction treatment, knowledges of the addict's faith base practice, and skills in addressing spiritual issues is important. Relying on a higher power does not mean merely telling the addict "don't worry God will cure you," rather, it means first helping the addict to see the consequences of his behavior, a conscious process that calls for genuine repentance that is not shaming or judgmental but rely on the Mercy of God who is not far away but resides within the addict to empower and liberate. The goal here is using motivational interviewing as earlier pointed out to elicit empathy and empowerment, helping the addict see and perceive in a new way that he/she has a higher power within that is able to liberate him/her from addiction.

Spirituality can also be used with the health belief model, another approach that promotes a biopsychosocial perspective. Health belief model focuses on the belief and attitude of the addict, and that changes occur when the individual has a perception of control. Even with those addicts who have no traditional religious or spiritual background, one can still tap into the broader view of spirituality as an inner power and/or beauty within that pulls humans toward finding meaning and purpose. This power within, is what dialectic behavioral treatment, another model that fits into bio-psychological approach, calls the "wise mind" within human that guide one to healthy choice and mastery.

In conclusion, the bio-psychosocial-spiritual approach is relevant as it describes the multifaceted process of addiction. It is noteworthy that it incorporated cultural domain to the social approach as culture is indeed complex and powerfully addresses the knowledge of a particularly group of people, their language, ethnicity, belief, traditions, and social habits; also that the culture of the addict sometimes determines if they will be light, moderate or heavy drinkers.

The reader does come to grip with the complex nature of addiction with a bio-psycho-social-spiritual approach that offers a more integral or integrated treatment to addiction or substance related disorders. The goal of a biopsychosocial-spiritual approach to addiction is well achieved and beneficial. It spurs the readers to conceptualize addiction treatment to include a multicultural perspective integrated into the interaction between biological, psychological, social and spiritual processes involved in the treatment of addictive behavior.

Franca Voke Niameh, PsyD
Psychologist & Clinical Supervisor
The Wright Institute's
Integrated Health Psychology Training Program (IHPTP)
Contra Costa Health Services, California, USA.

INTRODUCTION

Adeolu has been addicted to Methamphetamine and Alcohol since he was 18 years old. His addictions have led him to being incarcerated several times, and he never had stable jobs. He was diagnosed with bipolar disorder, he tested HIV positive, and has alcoholic liver cirrhosis. He has no stable relationships with families and friends as he has been very abusive. At the age of 45, he has 8 children from 6 women, none of whom he has seen in the last 10 years. He suffers depression, and he is angry towards God. Presently, he is homeless. Adeolu cries for help.

In recent times, there have been attention-shift in the approach to treatment of health related issues from just biological, psychological, social or spiritual approaches to holistic approach of an integrated biopsychosocial-spirituality. The same trend has occurred in the field of addiction that the past 20 years have witnessed various authors in the addiction field proposing biopsychosocial and spiritual model of addiction treatment.

1

Many factors, combined or interrelated, according to the World Health Organization, (2015), affect the health of individuals and communities; and their health is determined by their circumstances and environment which are visible through where we live, the state of our environment, genetics, our income and level of education, and our relationships with friends and family, and of course our beliefs and search for meaning. In other words, the determinants of health according to WHO (2015) include:

- the social and economic environment (which includes our religious and spiritual awareness and beliefs),
- the physical environment, and
- the person's individual characteristics and behaviors.

With this understanding, researchers are promoting biopsychosocial-spiritual model in response to the need for more complex, interactional and contextual paradigms which deviate from single-cause, linear models to multi-cause, interactive or integrated model (Schwartz, 1982; Kumpfer, K. L., Trunnell, E. P., Whiteside, H. O., 2015).

Understanding addiction from this framework of integrated biopsychosocial-spiritual factors and exploring the same factors during the intake and assessment of the patients/clients with addiction will lead to appropriate treatment plans and intervention which focus on the holistic treatment of the total person. Hence, the focus of the treatment will not only be on addiction but also taking care of the cause of addiction as well as redeeming what the addiction have damaged. In other words, the biopsychosocial-spiritual approach aims at working with the addicts into full functionality in order to fit back into the society just as some recovery programs now provide "assistance participants (who) need to help them transition back into society to be as productive as possible, within their family structure as the head of the household." (First Step Recovery Homes, Inc., 2012).

In treating Adeolu's addiction therefore, his medical, psychiatric, housing, family and spiritual situations will not be overlooked.

CHAPTER ONE

CONCEPT OF DRUG ADDICTION

There has been an age long attempt to classify mental and emotional disorders and illnesses that, in 1952 the American Psychiatric Association (APA) succeeded in publishing *Diagnostic and Statistical Manual of Mental Disorders (DSM)* whose fourth edition *DSM-IV-TR* divided "Substance-related disorders" into two general categories namely: Substance Use Disorders (SUDs) and Substance Induced Disorders (SIDs), (Inaba & Cohen, 2004). While Substance Use Disorders involve patterns of drug use which are divided into substance abuse and substance dependence, substance induced disorders refer to conditions that are caused by specific substances such as intoxication, withdrawal, and certain mental disorders (Inaba & Cohen).

In 2013, the *Diagnostic and Statistical Manual of Mental Disorders – Fifth Edition (DSM-5)* included a new chapter called "Substance-Related and Addictive Disorders" where substance use disorders (SUDs) are separately categorized with criteria that provide a gradation of severity within each diagnostic category. 'In so doing, the new edition has removed the terms "abuse" and "dependence" and has included the term *"addiction"* for the first time.' (Rogers Memorial Hospital, 2014).

What is Drug Addiction?

According to National Institute on Drug Abuse (NIDA), Drug Addiction is defined as:

> A chronic, relapsing brain disease that is characterized by compulsive drug seeking and use, despite harmful consequences. It is considered a brain disease because drugs change the brain; they change its structure and how it works. These brain changes can be long lasting and can lead to many harmful, often, self-destructive, behaviors.

Addiction to drug isn't a sudden or one time phenomenon. It is a brain disease that is initiated by behavior that begins from experimentation or accidental experience to abuse and finally to addiction. People use different drugs for different reasons. When an illegal drug is used or when a legal drug is used inappropriately, it is called an abuse which Darryl S. Inaba & William E. Cohen

(2004) represented as "the continued use of a drug despite negative consequences". Addiction occurs when the abuser cannot control the impulse to use the drug despite the negative effects (NIDA).

According to American Society of Addiction Medicine (ASAM, 2011), Addiction is characterized by "ABCDE":

a. Inability to consistently Abstain;
b. Impairment in Behavioral control;
c. Craving; or increased "hunger" for drugs or rewarding experiences;
d. Diminished recognition of significant problems with one's behaviors and interpersonal relationships; and
e. A dysfunctional Emotional response.

According to WHO (2015), "Global prevalence rates of drug use disorders among adults were estimated to range from 0% to 3% in 2004, with the highest prevalence rates found in the Eastern Mediterranean Region. The highest estimated prevalence rates of drug use disorders among men and women were found in parts of the Americas. Selected countries in Africa, Asia, the Eastern Mediterranean, Europe and the Western Pacific were found to have high rates of drug use disorders among men and women as well."

In the United States alone, the results of the national finding from the 2013 National Survey on Drug Use and Health summarily reveal the following as presented by SAMHSA (2013):

- In 2013, 39.0 percent of youths aged 12 to 17 perceived great risk in having five or more drinks once or twice a week. Similarly, 39.5 percent of youths perceived great risk in smoking marijuana once or twice a week.
- The percentage of youths aged 12 to 17 perceiving great risk in smoking marijuana once or twice a week decreased from 54.6 percent in 2007 to 39.5 percent in 2013.
- The percentage of youths who reported great risk in smoking one or more packs of cigarettes per day was 64.3 percent in 2013. The 2013 rate was lower than the rates between 2004 and 2009 (ranging from 65.5 to 69.5 percent) and was similar to the rates in 2002 (63.1 percent) and 2003 (64.2 percent).
- About half (48.6 percent) of youths aged 12 to 17 reported in 2013 that it would be "fairly easy" or "very easy" for them to obtain marijuana if they wanted some. One in eleven reported it would be

easy to get heroin (9.1 percent), 11.3 percent indicated that LSD would be easily available, and 14.4 percent reported easy availability for cocaine. In comparison with the rates in 2002, the 2013 rates represent declines in perceived availability for all four of these drugs.

- About one in eight youths aged 12 to 17 (12.4 percent) indicated that they had been approached by someone selling drugs in the past month, which was similar to the rate in 2012 (13.2 percent).

- A majority of youths aged 12 to 17 (88.4 percent) in 2013 reported that their parents would strongly disapprove of their trying marijuana once or twice, which was a decline from 2012 (89.3 percent). Current marijuana use was much less prevalent among youths who perceived strong parental disapproval for trying marijuana once or twice than for those who did not (4.1 vs. 29.3 percent, respectively).

- In 2013, 72.6 percent of youths aged 12 to 17 reported having seen or heard drug or alcohol prevention messages from sources outside of school, which was lower than in 2002 (83.2 percent) and in 2012 (75.9 percent). The percentage of school-enrolled youths reporting that they had seen or heard prevention messages at school also declined during this period, from 78.8 percent in 2002 to 73.5 percent. The prevalence of past month illicit drug use in 2013 was lower among youths who reported having such exposure to prevention messages compared with youths who did not have such exposure.

- In 2013, an estimated 21.6 million persons aged 12 or older (8.2 percent) were classified with substance dependence or abuse in the past year based on criteria specified in the Diagnostic and Statistical Manual of Mental Disorders, 4th edition (DSM-IV). Of these, 2.6 million were classified with dependence or abuse of both alcohol and illicit drugs, 4.3 million had dependence or abuse of illicit drugs but not alcohol, and 14.7 million had dependence or abuse of alcohol but not illicit drugs.

- The annual number of persons with substance dependence or abuse in 2013 (21.6 million) was similar to the number in each year from 2002 through 2012 (ranging from 20.6 million to 22.7 million).

- The specific illicit drugs with the largest numbers of persons with past year dependence or abuse in 2013 were marijuana (4.2 million), pain relievers (1.9 million), and cocaine (855,000). The number of persons with marijuana dependence or abuse was similar between 2002 and 2013. The number with pain reliever dependence or abuse in 2013 was similar to the numbers from 2006 to 2012. The number with cocaine dependence or abuse in 2013 was similar to the numbers in 2010 to 2012.

- The number of persons who had heroin dependence or abuse in 2013 (517,000) was similar to the numbers in 2009 to 2012 (ranging from 361,000 to 467,000), but it was higher than the numbers in 2002 to 2008 (ranging from 189,000 to 324,000).

- In 2013, adults aged 21 or older who had first used alcohol at age 14 or younger were more likely to be classified with alcohol dependence or abuse than adults who had their first drink at age 21 or older (14.8 vs. 2.3 percent).

- Between 2002 and 2013, the percentage of youths aged 12 to 17 with substance dependence or abuse declined from 8.9 to 5.2 percent. For young adults aged 18 to 25, substance dependence or abuse also declined during this period from 21.7 percent in 2002 to 17.3 percent in 2013.

- Treatment need is defined as having substance dependence or abuse or receiving substance use treatment at a specialty facility (hospital inpatient, drug or alcohol rehabilitation, or mental health centers) within the past 12 months. In 2013, 22.7 million persons aged 12 or older needed treatment for an illicit drug or alcohol use problem (8.6 percent of persons aged 12 or older). Of these, 2.5 million (0.9 percent of persons aged 12 or older and 10.9 percent of those who needed treatment) received treatment at a specialty facility. Thus, 20.2 million persons (7.7 percent of the population aged 12 or older) needed treatment for an illicit drug or alcohol use problem but did not receive treatment at a specialty facility in the past year.

- Of the 20.2 million persons aged 12 or older in 2013 who were classified as needing substance use treatment but did not receive treatment at a specialty facility in the past year, 908,000 persons (4.5 percent) reported that they felt they needed treatment for their illicit drug or alcohol use problem. Of these 908,000 persons who felt they needed treatment, 316,000 (34.8 percent) reported that they made an effort to get treatment. Based on combined 2010-

2013 data, the most commonly reported reason for not receiving treatment among this group of persons was a lack of insurance coverage and inability to afford the cost (37.3 percent).

Also, according to the National Drug Intelligence Center, (2011), drug abuse and addiction to illicit and prescription substances cost more than $700 billion a year in increased health care costs, crime, and loss of productivity. And every year, the use of alcohol and illicit and prescription drugs contribute to more than 90,000 deaths of Americans, while tobacco alone is linked to death of about 480,000 people every year, according to Center for Disease Control and Prevention (CDC).

In order to fully understand why people become addicted to alcohol and/or different drugs of choice some theories have been developed which suggest genetic and other biological factors as the causes of addiction, while some others emphasize personality factors or social-environmental factors (Lettieri et al., 1980; Ogborne, A., 2015). And while these factors contribute to persistent substance use, addiction, and to relapse following periods of abstinence, no one set of factors can account for all types of substance use. Substance use involve complex interactions of biological, psychological and social-environmental structures and processes (Arif & Westermeyer, 1988; Ogborne, A., 2015), and spirituality (Alternatives In Treatment, 2014). Hence, the American Society of Addiction Medicine (ASAM, 2011), opined that there are other factors that contribute to the appearance of addiction, which lead to its characteristic bio-psycho-socio-spiritual manifestations. These factors include:

a. The presence of an underlying biological deficit in the function of reward circuits, such that drugs and behaviors which enhance reward function are preferred and sought as reinforcers;

b. The repeated engagement in drug use or other addictive behaviors, causing neuroadaptation in motivational circuitry leading to impaired control over further drug use or engagement in addictive behaviors;

c. Cognitive and affective distortions, which impair perceptions and compromise the ability to deal with feelings, resulting in significant self-deception;

d. Disruption of healthy social supports and problems in interpersonal relationships which impact the development or impact of resiliencies;

e. Exposure to trauma or stressors that overwhelm an individual's coping abilities;

f. Distortion in meaning, purpose and values that guide attitudes, thinking and behavior;

g. Distortions in a person's connection with self, with others and with the transcendent (referred to as God by many, the Higher Power by 12-steps groups, or higher consciousness by others); and

h. The presence of co-occurring psychiatric disorders in persons who engage in substance use or other addictive behaviors.

Another characteristic of addiction involves the power of external cues which is capable to trigger craving and drug use, as well as increasing the frequency of perpetuating addictive behaviors (ASAM, 2011). This involves the hippocampus which is important in memory of previous euphoric or dysphoric experiences, and the amygdala which is important in having motivation concentrating on selecting behaviors associated with these past experiences (ASAM, 2011). Although quantity or frequency of drug use may make the difference between those who have addiction, the qualitative way in which an individual respond to external cues also matters. In fact, "a particularly pathological aspect of *the way* that persons with addiction pursue substance use or external rewards is that preoccupation with, obsession with and/or pursuit of rewards (e.g., alcohol and other drug use) persist despite the accumulation of adverse consequences. These manifestations can occur compulsively or impulsively, as a reflection of impaired control" (ASAM, 2011).

CHAPTER TWO

CONCEPT OF BIOPSYCHOSOCIAL-SPIRITUAL APPROACH

Science and medicine have gone a long way into treating human diseases; and health professions have become outstanding at looking into the physical finitude of human body and cured it of diseases (Sulmacy, D. P, 2002) or scientifically support, replace or supplement human parts. It has equally been proven that psychosocial factors affect the onset and cause of almost all chronic physical disorders, and that mechanisms like neuroendocrine and immunological defense may regulate the effects of psychological factors on physical process (Dogar, I. A., 2007). Yet, neither contemporary medicine nor social psychology has adequately addressed the complex nature of human disease as a whole. "The contemporary medicine still stands justly accused of having failed to address itself to the needs of whole human persons and of preferring to limit its attention to the finitude of human bodies (Ramsey, 1970; Sulmasy, D. P., 2002).

According to the philosopher Karl Max, (1976), "Nothing in the world is more complex or more perplexing than a human being". Human being is an embodiment of the interaction between genetics and environment, hence, to understand human beings, one has to put into consideration human biological, psychological, social and spiritual factors; and the healing professions need also to put these factors into perspective. In other words, "the genuinely health care must address the totality of the patient's relational existence-physical, psychological, social, and spiritual" (Sulmasy, D.P., 2002)

In other to take these factors into consideration, "George Engel proposed the biopsychosocial model in what soon became a landmark event for understanding medicine as a science" (Smith, R.C., 2002). By doing so, making the biopsychosocial model a new, broader, and integrated approach to human behavior and disease (Dogar, I. A., 2007). Although, George Engel did not include spirituality, the fact that many psychotherapeutic clients are finding out that the incorporation of spiritual awareness and understanding are effective adjuncts to their psychological treatments (Garfolo, B., 2015), made some to call for a model that goes even further – a *biopsychosocial-spiritual model* of healthcare (King, 2000; McKee & Chappel., 1992; Sulmasy, D.P., 2002).

Worth noting however, that both biopsychosocial model and biopsychosocial-spiritual model point to the same approach of looking at human behavior and approaching treatment of human diseases.

With the growth in addiction research over the years, "some researchers have concluded that addiction affects the body, the mind, and the spirit. Along with treatment tools based on psychological, physical, and social needs, spirituality has long played a role in recovery from addiction." (Alternatives In Treatment, 2014). The biopsychosocial-spiritual (BPSS) model have been very useful for some addiction professionals in the attempt to answer the question, "How do people get addicted?" (Horvath, A. T., et al, 2015). Although, some people feel it is of no importance to separately mention or adding spirituality to biopsychosocial model, nonetheless, spirituality is a very important reality for some other people.

Hence, Horvath, A.T, ABPP, Misra, K., Epner, A. K., & Cooper, G.M., (2015) opined that,

> Adding "Spirituality" to the Bio-Psycho-Social model assists some people to move beyond the physicality of their addiction. By adding spirituality into recovery efforts, many new healing possibilities become available. For instance, it is possible to envision addiction as a loss of one's humanity. Our true, authentic, and spiritual self has become disconnected from our physical being because of the addiction. Therefore, a change that reunites the authentic spiritual self, with the physical body, would be healing. Alternatively, it might be possible to understand addiction as a way of coping with a previous loss of our true authentic self. This kind of loss might occur from trauma such as abuse. These kinds of trauma often shatter our belief in a meaning and purpose to life.

Biopsychosocial model or biopsychosocial-spiritual model "offers an integrated conceptual framework useful to all professionals and researchers in the area of prevention and treatment of addiction (Kumpfer, K. L., Trunnell, F. P., & Whiteside, H. O., (2015).

(A)

BIOLOGICAL UNDERSTANDING OF ADDICTION

The question that remains motivational to searching into the understanding of addiction is "why do some people get addicted to a particular drug or behavior and some other people not"? From the biological perspective, the answer hinges on the genetics and neurological studies which show that addiction can be inherited and at the same time seen as the disease of the brain.

Genetics factor of Addiction.

Geneticists, for years, have discovered that many traits such as eye color, bone structures, and of course, the chemistry of the nervous system are passed from generations to generations by genes (Inaba & Cohen, 2004). And as geneticists progress in understanding the role of genetics in various conditions and diseases, it was also discovered that there is likely to be a genetic component to substance abuse and addiction. Meaning that, inherited differences among individuals affect their reactions to drugs (https://www.princeton.edu/~ota/disk1/1993/9311/9311 06.PDF). However, it is good to note that "there isn't just one gene that affects addiction. There are more than 100 that have been associated with drug abuse" (Inaba & Cohen). And according to *Learn Genetics*, a University of Utah Health Sciences online publication,

> Scientists will never find just one single addiction gene. Like most other diseases, addiction vulnerability is a very complex trait. Many factors determine the likelihood that someone will become an addict, including both inherited and environmental factors.
>
> Because addiction is a complex disease, finding addiction genes can be a tricky process. Multiple genes and environmental factors can add up to make an individual susceptible, or they may cancel each other out. Not every addict will carry the same gene, and not everyone who carries an addiction gene will exhibit the trait. (http://learn.genetics.utah.edu/content/addict ion/genes/)

Yet, "Genetic factors account for about half of the likelihood that an individual will develop addiction" (ASAM, 2011) that scientists have estimated that genetic factors account for between 40 and 60 percent of a person's vulnerability to addiction which includes the effects of environmental factors on the function and expression of a person's genes. A person's stage of development and other medical conditions also contribute to vulnerability to addiction (NIDA, July 2014).

Many studies have been carried out, in this regard, to discover the genes that play roles in addiction such as mice genetic studies, family studies, twins studies etc.

Mice Studies

Mice have been very useful in the animal models in identifying addiction genes simply because the reward pathway as well as many of the genes that underlie it, function similar to the way it functions in human beings. When an addiction gene is identified in mice, its counterpart gene can be identified in humans by looking for similar DNA sequences. As presented in the *Learn Genetics*, the following were found out from mice studies:

- The A1 allele of the dopamine receptor gene *DRD2* is more common in people addicted to alcohol or cocaine.
- with increased expression of the *Mpdz* gene experience less severe withdrawal symptoms from sedative-hypnotic drugs such as barbiturates.

- Mice without the cannabinoid receptor gene *Cnr1* are less responsive to morphine.
- Mice lacking the serotonin receptor gene *Htr1b* are more attracted to cocaine and alcohol.
- Mice bred to lack the $\beta 2$ subunit of nicotinic cholinergic receptors have a reduced reward response to cocaine.
- Mice with low levels of neuropeptide Y drink more alcohol, whereas those with higher levels tend to abstain.
- Mice mutated with a defective *Per2* gene drink three times more alcohol than normal.
- Non-smokers are more likely than smokers to carry a protective allele of the *CYP2A6* gene, which causes them to feel nausea and dizziness from smoking.
- Alcoholism is rare in people with two copies of the *ALDH*2* gene variation.
- Mice lacking the *Creb* gene are less likely to develop morphine dependence.

(http://learn.genetics.utah.edu/content/addiction/genes/)

Biological Family Studies of Alcoholics

It is no myth that addiction runs in families, just as the recent years' studies have revealed that physical reactions and diseases or disorders such as some form of Alzheimer's disease, schizophrenia, some form of depression, juvenile diabetes are genetically transmittable (Inaba & Cohen). The studies of the families of alcoholics are examples. Although, "while family studies can establish that a disorder (or liability to a disorder) is transmitted; in general, they are unable to distinguish between biological and cultural transmission" (https://www.princeton.edu/~ota/disk1/1993/9311/931106.PDF).

Reviewing the various biologic family records of alcoholics in various treatment programs across the United States, for example, it was revealed that:

> If one biological parent was alcoholic, a male child was about 34% more likely to be an alcoholic that the male child of non-alcoholics. If both biological parents were alcoholic, the child was about 400% more likely to be alcoholic. If both parents and a grandfather were alcoholics, that child was about 900% more likely to develop alcoholism. About 28 million Americans

have at least one alcoholic parent (Schuckit, 1986, Inaba & Cohen, 2004)

The family trees below give examples of families with alcoholism. The square shape represent male while the circle shape represent women. The individuals in shaded square or circle are affected by alcoholism while those in plane square or circle are unaffected. The pedigree or family tree below was taken from the Genetic Science Learning Center, University of Utah.

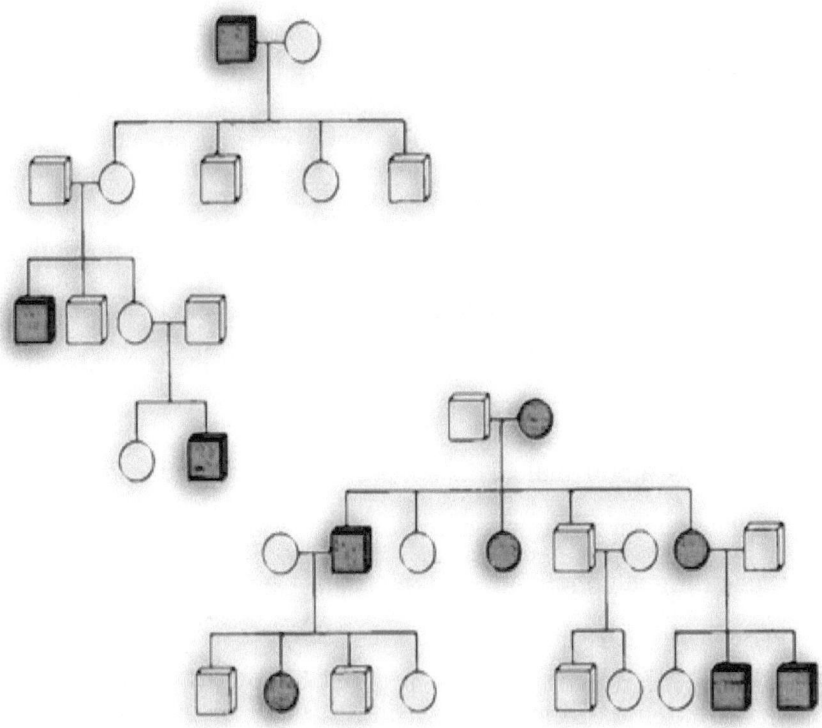

Twins Studies

Twin studies have offered a critical method to studying questions of nature and nurture, as well as the interaction between genetics and environment in addiction research. For according to Dr. Naimah Weinberg of NIDA's Division of Epidemiology, "Twin studies explore the roles and interrelationship of genetic and environmental risk factors in the development of drug use, abuse, and dependence" (Zickler, P., 1999).

The Scientists at the Virginia Institute for Psychiatric and Behavioral Genetics (VIPBG) at Virginia Commonwealth University use data from thousands of twins all over the world which include data from pairs of twins from Virginia, Sweden, Finland and China, acquired through VCU collaborations (Beck, T., 2013). Identical and fraternal twins were studied. "The Virginia Adult Twin Study of Psychiatric & Substance Use Disorder (VATSPSUD) includes 4,500 male and female twin pairs and examined seven common psychiatric and addiction disorders, including alcoholism, and nicotine and drug addiction (Beck, T., 2013).

The Virginia Twin Studies showed that "in early adolescence the initiation and use of nicotine, alcohol, and cannabis are more strongly determined by familial and social factors, but these gradually decline in importance during the progression to young and middle adulthood, when the effects of genetic factors become maximal, declining somewhat with aging" (Bevilacqua L. & Godlman, D., 2009)

In another research carried out by Dr. Kenneth Kendler and Dr. Carol Prescott at the Medical College of Virginia in Richmond aimed at examining the patterns of marijuana and cocaine use by female twins, it was found that genetic factors play a major role in the progression from drug use to abuse and dependence. (Zickler, P., 1999). In an interview of 1,934 twins from age 22 to 62, recruited from the Virginia Twin Registry, a database compiled from Commonwealth birth records, Dr. Kendler says, "our findings suggest that the progression from the use of cocaine or marijuana to abuse or dependence was due largely to genetic factors." Also, the study found that concordance rates-both twins using, abusing, or being dependent on drugs-were higher for identical than fraternal twins (Zickler, P., 1999). For instance, "for cocaine use, concordance was 54 percent in identical twins and 42 percent in fraternal twins; for abuse, 47 percent in identical twins and 8 percent in fraternal twins; and for dependence, 35 percent in identical twins and zero for fraternal twins" (Zickler, P., 1999).

Disease model of Addiction

The strongest claim in the Biological understanding of addiction hinges around the consideration and classification of alcoholism and drug addiction as one of such diseases; such that, like heart disease, high blood pressure, diabetes, Alzheimer and asthma, addiction has specific risk factors which, if not effectively treated, can lead to other illnesses and even death (CASAColumbia, 2012). According to Inaba and Cohen (2004), much of the current research in the treatment of alcoholism, for example, is based on the concept of alcoholism as a disease – an idea which can be dated back to thousands of years ago and which only recently become accepted. Hence, alcoholism is conceived as a "chronic progressive disease" by the National Council on Alcoholism in 1972 (Inaba and Cohen) while the medical community believe that drug addiction in general is a treatable brain disease. The American Society of Addiction Medicine (ASAM, 2011) therefore defines and describes *addiction* as

A primary, chronic disease of brain reward, motivation, memory and related circuitry. Addiction affects neurotransmission and interactions within reward structures of the brain, including the nucleus accumbens, anterior cingulate cortex, basal forebrain and amygdala, such that motivational hierarchies are altered and addictive behaviors, which may or may not include alcohol and other drug use, supplant healthy, self-care related behaviors. Addiction also affects neurotransmission and interactions between cortical and hippocampal circuits and brain reward structures, such that the memory of previous exposures to rewards (such as food, sex, alcohol and other drugs) leads to a biological and behavioral response to external cues, in turn triggering craving and/or engagement in addictive behaviors. It doesn't matter how drugs are taken – inhaled, injected, snorted, drunk, or absorbed by contact – the target is the brain, as it is being transported through the blood stream. When the drug gets to the brain, it alters brain's structure and functioning causing the person with this disease to become addicted to the drug (Horvath, A.T, ABPP, Misra, K., Epner, A. K., & Cooper, G.M.; ed. Zupanick, C. E., 2015). The neurobiology of addiction involves more than the neurochemistry of reward.

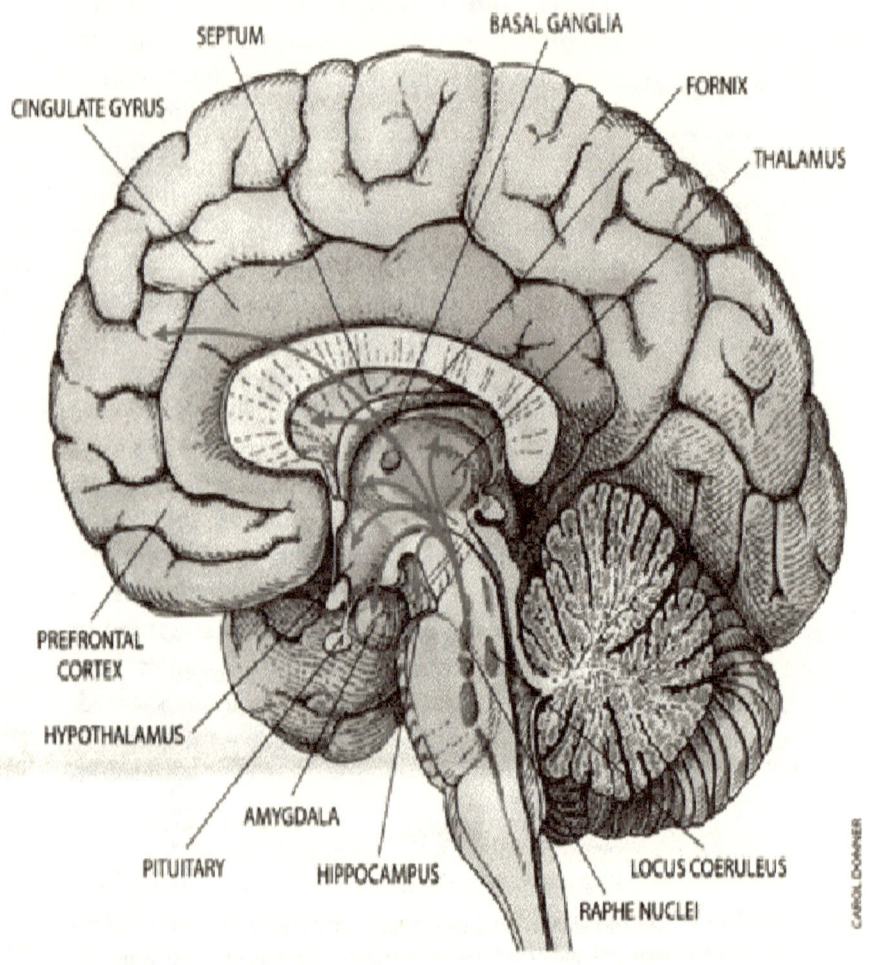

SEPTUM

BASAL GANGLIA

FORNIX

THALAMUS

CINGULATE GYRUS

PREFRONTAL
CORTEX

HYPOTHALAMUS

AMYGDALA

PITUITARY

HIPPOCAMPUS

LOCUS COERULEUS

RAPHE NUCLEI

CAROL DONNER

(Diagram from ttp://faculty.uca.edu/lray/brain.html)

The frontal cortex of the brain and underlying white matter connections between the frontal cortex and circuits of reward, motivation and memory are fundamental in the manifestations of altered impulse control, altered judgment, and the dysfunctional pursuit of rewards (which is often experienced by the affected person as a desire to "be normal") seen in addiction--despite cumulative adverse consequences experienced from engagement in substance use and other addictive behaviors. The frontal lobes are important in inhibiting impulsivity and in assisting individuals to appropriately delay gratification. When persons with addiction manifest problems in deferring gratification, there is a neurological locus of these problems in the frontal cortex. (ASAM, 2011)

Meaning that a long term use of drug(s) of choice can affect judgement and behavior which in some cases can lead to compulsive desire to obtain and use same drug(s) despite the negative or harmful or dangerous consequence resulting from taking the drug. The changes in the brain can remain even after the person stops using the drug(s) (CASAColumbia, 2012)

The brain disease model of addiction is strongly supported by scientific evidence, as recognized by the NIDA Director, Dr. Nora Volkow, and NIAAA Director, Dr. George Koob. They claimed that animal and human studies have shown that critical brain structures and behaviors are disrupted by chronic exposure to drugs and alcohol (NIDA,

July 29, 2015).   For them, "These findings, along with ongoing research, are helping to explain how drugs and alcohol affect brain processes associated with loss of control, compulsive drug taking, inflexible behavior, and negative emotional states associated with addiction" (NIDA, July 29, 2015).   Also the understanding of these processes has resulted in several effective medicines, as well as new and promising medication which is targeted to treat drug and alcohol addiction (NIDA, July 29, 2015). This "process of discovery within a disease framework has also led to the development of promising brain stimulation treatments and behavioral interventions, and has had a positive impact on public policy" (NIDA, July 29, 2015).

(B)

Psychological Understanding of Addiction

The American Psychological Association (APA)'s glossary of psychological terms by Gerrig, Richard J. & Philip G. Zimbardo, presents *Psychology* as "The scientific study of the behavior of individuals and their mental processes". Meaning that psychologists are concerned with improving the quality of people's lives as well as their life satisfaction by considering behaviors that promote people's well-being and life satisfaction adaptive behaviors which limit people's functioning and diminish life satisfaction (Horvath, A.T, ABPP, Misra, K., Epner, A. K., & Cooper, G.M., 2015). Any behaviors which limit people's functioning and/or diminish life satisfaction is what is called maladaptive behaviors (Horvath, A.T, ABPP, Misra, K., Epner, A. K., & Cooper, G.M., 2015). Addiction is one of such behaviors.

Historically, the psychological understanding of addiction can be traced back to the perspective of Sigmund Freud who would think of addiction as a subconscious processes. According to Dr Gary S. Fisher (2010),

Freud saw all addictions (chemical, behavioral, etc.) as substitutes for primitive sexual urges which created an internal conflict and increased guilt which is reenacted through the addictive cycle. Since Freud, most of the major psychoanalytic-based theories of addiction continue to view it as the result of other deeply buried conflicts sometimes also related to sexual issues, or an acting out of aggression or rage, as well as ways of coping with fears, dependency needs, and countering feelings of powerlessness.

Transactional Analysis expands on Freud's concept of the ego in order to explain mental issues and problems in "transactions" with others. It basically focuses on the concept of life scripts and choices in regards to how the individual makes a decision "to lead a self- destructive life based on a game of 'Alcoholic'" (Steiner, 1971, p.xvii; Fisher, G., 2010)

Dr. Gary Fisher continues to unveil the psychological understanding of why people become addicted stating that,

Other psychodynamic approaches attempt to address the addictive personality and those traits associated with it; immaturity, grandiosity, low self-esteem, and an unwillingness to face reality (Kissin, 1977).

Self-psychology attempts to incorporate most of the other psychodynamic theories of addiction (especially psychoanalytic) under the umbrella of addiction as a disturbance of the experience of the self, and in particular, that the individual is stuck in the archaic self, an immature stage in which the individual suffers from narcissism and can't work out of it.

The National Institute on Drug Abuse relates that, "Cognitive-behavioral strategies are based on the theory that in the development of maladaptive behavioral patterns like substance abuse, learning processes play a critical role." This is because, the Cognitive Behavioral Therapy (CBT) "views addiction as a coping mechanism developed to cope with stirred up emotions created by negative thought patterns" (Fisher, G., 2010).

Helen Blair Simpson, M.D., Ph.D., a Professor of Psychiatry at Columbia University Medical Center and Director of Center for Obsessive-Compulsive and Related Disorders however stated that, "There's a new model of addictions which says it starts with you impulsively taking a substance but then as you take the substance in fact you lose the ability to control the ability to take it anymore and that addiction becomes a compulsion ..." That is, addiction can be compared with, and as it is related to, other psychological disorders like Obsessive Compulsive Disorders (OCD). Take for instance, alcohol and other drugs of abuse may temporarily relieve the anxiety and depression associated with OCD, but unfortunately, these substances may make OCD symptoms worse. Drug and alcohol abuse seem to be common in people who suffer anxiety disorders including OCD. Using substance of abuse to self-medicate can lead to other health problems, financial troubles, legal conflicts, unemployment and addiction.

(http://www.futuresofpalmbeach.com/co-occurring-disorders-overview/ocd/)

On the other hand, disorders of compulsion, such as OCD and Tourette's, are thought to be distinct from disorders of addiction with the main difference been that while addicts have the ability to abstain from satisfying the addiction, individuals with compulsive disorders have no such ability (Marlow K, 2013).

As it is known that some of the most common mental health disorders found in chemically dependent people include mood- and anxiety disorders, an even higher percentage of people with severe mental illness also have co-occurring substance use disorders. Called severe because of the severity and length of episodes of illness, these mental health disorders include schizophrenia and schizoaffective disorder (http://www.bhevolution.org/public/cooccurring_overview.page).

Though, it can sometimes be difficult to diagnose Co-occurring disorders because symptoms of addiction can mask symptoms of mental illness, and the symptoms of mental illness can be muddled up with symptoms of addiction (http://www.bhevolution.org/public/cooccurring_overview.page). The truth is that, as found in the DualDiagnosis.org (http://www.dualdiagnosis.org/co-occurring-disorders/),

- Almost 9 million men and women who abuse drugs or alcohol have a mental health issue, also known as a co-occurring disorder or a Dual Diagnosis.
- Out of all of the adults who go through addiction treatment, only about 7 percent are treated for both their substance abuse and their co-occurring disorder.
- Over 55 percent of those who suffer from a co-occurring disorder get any help at all.
- The rate of homelessness among people with co-occurring disorders is approximately 23 percent.

Also, SAMHSA's *The Evidence-ITC,* (2009), revealed that Client with dual diagnosis or co-occurring disorders will therefore require treatments that will combine or integrate mental health and substance abuse interventions at the level of the clinical interaction.

> In other words, the caregivers take responsibility for combining the interventions into one coherent package. For the individual with a dual diagnosis, the services appear seamless, with a consistent approach, philosophy, and set of recommendations. The need to negotiate with separate clinical teams, programs, or systems disappears. Integration involves not only combining appropriate treatments for both disorders but also modifying traditional interventions For example, social skills training emphasizes the importance of developing relationships but also the need to avoid social situations that could lead to substance use. Substance abuse counseling goes slowly, in accordance with the cognitive deficits, negative symptoms, vulnerability to confrontation, and greater need for support that are characteristic of many individuals with severe mental illness. Family interventions address understanding and learning to cope with two interacting illnesses.

(C)

Sociological Understanding of Addiction

In other to understand addiction it is important also to note that sociologists utilize a somewhat different approach in defining drug use and abuse than scholars from the fields of biology, pharmacology, and psychology by focusing more on the social meaning of drugs and alcohol, norms and patterns regarding their consumption in certain settings, and consequences resulting therefrom, rather than focusing on genetic predispositions, chemical imbalances, neurological processes, or personality traits.

(http://www.udel.edu/soc/tammya/pdf/crju369_theory.pdf)

Addiction is not just a psychological factor which deals with individual behavior or biological factor of a disease model, it is also cause by the sociological circumstances and as well affect the same social structure the addict may find him/herself. In other words, the society and the different social groups such as families, organizations, and peers can contribute and be affected by a person's addiction that one of the ways to understand and correct addiction is within the context of the groups in which it occurs (Horvath, A.T, ABPP, Misra, K., Epner, A. K., & Cooper, G.M.; Ed. Zupanick, C. E., August 26, 2013).

Our environment surely has impact on the persons' addiction just as addiction has impact on our environment. A person's environment according to National Institute on Drug Abuse "includes many different influences, from family and friends to socioeconomic status and quality of life in general. Factors such as peer pressure, physical and sexual abuse, stress, and quality of parenting can greatly influence the occurrence of drug abuse and the escalation to addiction in a person's life." (NIDA, 2012)

Although there are many sociological theories of addiction, the earliest most influential sociological research on addiction was conducted by Alfred Lindesmith (1938, 1940, 1947, 1968), and his theory has remained a classic sociological theory of addiction (Akers 1992; McAuliffe and Gordon, 1974; Stephens, 1991; Weinberg, 1997a, 1998; Weinberg, D., 2011). According to Lindesmith, in order to understand addiction it is necessary to consider the addicts' subjective perception of drugs, drug effects, and their wider social lives (Weinberg, D. 2011)

So, to fully get to understand how people get addicted, one need to understand the social and cultural forces. The three primary socio-cultural influences to be examined are culture, families and social support (Horvath, A.T, ABPP, Misra, K., Epner, A. K., & Cooper, G.M.; Ed. Zupanick, C. E., August 26, 2013).

Culture is a very complex concept with a powerful influence on how people behave. It can be understood not only from the beliefs and attitudes a particular group of people, but also from the race and ethnicity, social status, level or education, geographical location, religion, legal status, etc. There are some cultures which encourage moderate use of alcohol, while some breed heavy drinking.

> For instance, Italian, Spanish, French, Greek, Jewish and Chinese do not usually have significant alcohol problems. In these cultures, drinking does not typically occur for the sake of getting high. Rather, it occurs in the context of a meal, ritual, or celebration. Or, do you belong to one of the heavy-drinking cultures where alcohol abuse, and/or other drug abuse, is more the norm? This includes the cultures and sub-cultures of Russia, Ireland, Scotland, and various US college campuses and fraternities, to name just a few. (Horvath, A.T, ABPP, Misra, K., Epner, A. K., & Cooper, G.M.; Ed. Zupanick, C. E., August 26, 2013).

The Sober College also put together the differential cultural factors:

> An impoverished environment can increase the likelihood of abuse. Poverty can affect generations of family members due to lack of education and limited access to employment or healthcare. Poverty-stricken environments leave many experiencing lifestyles including incarceration, homelessness and poor health. Those who drop out of school, are unemployed or live in unsafe areas are at higher risk, especially if their home environment has already exposed them to drugs and/or alcohol. (http://sobercollege.com/contributing-factors-substance-abuse-addiction/)

Another sociological influence is society's social sanctions, and the severity of those sanctions which may include legal sanctions for alcohol or drug use such as, punishment, and social stigma. Hence, the mid-twentieth century witnessed the input of functionalist sociologists with their various theories of addictive behavior. They consider addiction not as a loss of self-control as some believe rather as a result of deviant use of drugs (Weinberg, D. 2011). Robert Merton (1938) in his famous essay, "Social Structure and Anomie" suggested that "chronic drunkard and drug addicts might exemplify the retreatist adaptation, one of his five modes of adjustment whereby people adopt ostensibly deviant patterns of action" (Weinberg, D. 2011). Meaning that addicts are persons who believe in the respectability of the cultural goals and the institutionalized procedures that the society offers and yet cannot achieve these through sanctions; "the result is a retreat from social life into 'defeatism, quietism, and resignation'" (Merton, 1938, 678; Weinberg, D., 2011)

Families influence their members, particularly children, in multiple ways in their choices about smoking, drinking and using other drugs. According to the National Council on Alcoholism and Drug Dependence, Incorporation (NCADD), the number one health problem in the United State is alcoholism and drug dependence and they affect not only the individual, but millions of family members. In fact, "the disease of alcoholism and addiction is a family disease and affects everyone close to the person (NCADD). "Fathers, mothers, single parents, couples straight or gay, regardless of ethnicity or social group, rich or poor... drug and alcohol abuse can destroy relationships. Most of all, young children and adolescents suffer the greatest from the effects of the abuse of alcohol and drugs in the family" (NCADD).

In April 2004, the National Center on Addiction and Substance Abuse (CASA) at Columbia University hosted a CASACONFERENCE with the theme: *Family Matters: Substance Abuse and The American Family.* The conference examined the following:

1. Situations and characteristics that influence children's risk of abusing substances
2. What parents can do to reduce their children's risk of substance abuse
3. How parents can spot substance abuse by their children and what to do when they spot it
4. The impact of substance abuse on the American family, in relation to divorce, teen pregnancy, child and spousal abuse, and juvenile delinquency

It was found out that "Parents who used tobacco or illegal drugs or abused alcohol put half the nation's children—more than 35 million of them—at greater risk of substance abuse and other physical and mental illnesses. The report found that of all children under age 18:

- 13% lived in a household where a parent or other adult used illicit drugs
- 24% lived in a household where a parent or other adult was a binge or heavy drinker

- 37% lived in a household where a parent or other adult smoked or chewed tobacco" (CASAColumbia, 2005)

In the same vein, the National Institute on Drug Abuse stated that addiction runs in families such that if one's one or both parents, for example, are addicts, one is most likely to become addicted to drugs if used; just like one having a greater chance of having heart disease because one's parent or many of the relatives are having it. "Often many people in a family will have drug problems. It can be a problem that continues through many generations. This can happen whether the family is rich, poor, or in between." (NIDA)

This great possibility is obvious when parents are addicted to alcohol or illegal drugs. Life at home can be very stressful and unhappy for the children and of course the family can be dysfunctional. The result or the effect being the children themselves becoming addicts when they grow up (NIDA). Dr. Tian Dayton, puts this beautifully in her article *Living with Addiction: What happens to the family when addiction becomes part of it?*

Families where addiction is present are oftentimes painful to live in, which is why those who live with addiction may become traumatized to varying degrees by the experience. Broad swings, from one end of the emotional, psychological and behavioral spectrum to the other, all too often characterize the addicted family system. Living with addiction can put family members under unusual stress. Normal routines are constantly being interrupted by unexpected or even frightening kinds of experiences that are part of living with drug use. What is being said often doesn't match up with what family members sense, feel beneath the surface or see right in front of their eyes. The drug user as well as family members may bend, manipulate and deny reality in their attempt to maintain a family order that they experience as gradually slipping away. The entire system becomes absorbed by a problem that is slowly spinning out of control. Little things become big and big things get minimized as pain is denied and slips out sideways. (http://www.nacoa.org/pdfs/The%20Set%20Up%20for%20Social%20Work%20Curriculum.pdf)

However, "the good news is that many children whose parents had drug problems do not become addicted when they grow up. The risk is higher but it does not have to happen. And you can protect

yourself from the risk by not abusing drugs at all",
says National Institute on Drug Abuse.

(D)

Spiritual Understanding of Addiction

One thing that most scholars in the field of addiction try to first clarify is the connection between spirituality and religion. For many, spirituality and religion are two different things. Therefore, many try to separate spirituality and religion even though they are unconsciously linked together.

> We often unconsciously link the two. But spirituality does not need to be defined through the lens of religion. Religion can be thought of as a set of beliefs, rituals and practices regarding belief in God or gods to be worshipped. Spirituality is a personal search for meaning in life, for connection with all things and for the experience of a power beyond oneself. Some find it helpful to think of religion as rules or practices agreed to by a number of people, whereas spirituality is completely related to one's individual experience and connections. Spirituality is recognizing a power greater than ourselves which is grounded in love and compassion. It is a power that gives us perspective, meaning, and a purpose to our lives. It is a desire to connect with more than ourselves, to connect with everything. (http://www.new-hope-recovery.com/center/2014/07/10/spirituality-important-addiction-recovery/)

In fact, there are so many perspectives and understanding of the term "spirituality" that it has become a personal thing. Tim Stoddart (2013) thus refer to "Spirituality" as a vague term, which notwithstanding means a lot especially for addicts and recovering addicts.

But it will be absurd to out-rightly eliminate religion from the understanding of spirituality for different religions have their own spirituality such as Christian spirituality, Buddhist spirituality, Islamic spirituality etc; and the members of the different religions practice their spirituality as it connects to the spirituality of their different religions. This is expressed in their connection to their religious beliefs and values as well as morality. We all have a sense of morality that the immoral people even have morality of sorts; and our morality may come from our belief in a higher power or from deeply-held values (Horvath, A.T, ABPP, Misra, K., Epner, A. K., & Cooper, G.M.; ed. Zupanick, C. E., 2015).

Relationship between spirituality and addiction

According to Sobriety Coach, Cynthia Perkins, (2015), the relationship between addiction and spirituality is complex and multi-faceted, and not clearly understood by the masses. For on one hand, if one is empty spiritually there is the possibility to seek out artificial means to fill the gap, and on the other hand addiction itself leaves one empty on the spiritual level. In fact, a disconnection from God or a higher power causes addiction (Horvath, A.T, ABPP, Misra, K., Epner, A. K., & Cooper, G.M.; ed. Zupanick, C. E., 2015). When one is disconnected, one may fail to live according to God's will or direction and invariably may fail morally, to the extent of one having the feeling of emptiness and hopelessness that may result in addiction to a substance which pretends to fill this lacuna.

Spirituality greatly influence the use of alcohol and substance. Research shows that being a part of "a faith-based community, participating in religious activities, or associating with a network of individuals who share similar beliefs increase self-esteem, wellbeing, and a feeling of belonging" (Alternative in Treatment, 2014). Meaning that religious beliefs and practices impact addiction. There are three potential ways according to *Alternative in Treatment* (2014) in which this is expressed:

1. A member of a certain religion may be less likely to engage in illicit drug use if that religion prohibits it

2. Participation in a religious or like-minded community may fill social voids, providing both a sense of belonging and acceptance

3. Connection or closeness to a higher power provides a feeling of optimism for the future, as well as strength to resist substance use

Maria Mooney (February 26, 2012) opined that, when an individual uses his/her alcohol or substance of choice, what happens is a detachment and disconnection from the present moment, an uncomfortable feeling that one seeks to avoid through self-medication, and ultimately, to avoid the self. This is because "Addiction is a disease of isolation, and as the individual sinks deeper and deeper into the disease, he/she becomes more isolated from others and oneself as deeply rooted feelings of inner insufficiency and not being 'enough' create the overwhelming need to use" (Mooney, M., 2012). As the disease of addiction grabs hold of your body, it begins to impact one's spirituality or spiritual life (Stoddart, T., May 8, 2013). Addiction then gradually damages one physically, one begins to break down spiritually, too; resulting into lying to others about the use of drug or alcohol, and as well lying to oneself (Stoddart, T., May 8, 2013). "Eventually, you will forego your morals in order to get your fix, stealing and hurting others when necessary. You become a person who you and your loved ones don't even recognize, because you are saying and doing things that you would've considered morally wrong in the past when you were sober." (Stoddart, T., May 8, 2013.

In fact, a lack of connection to authentic self, important others, a higher power, and the larger community can contribute to the feelings of isolation and emptiness, low self-worth, and a pervasive sense of unhappiness that can lead to or result into perpetuating addictive behaviors (Mooney, M., 2012). So, "being of service is a profound way that recovering individuals often give back and regain a sense of self-worth and purpose as they work toward maintaining long-term sobriety" (Mooney, M., 2012).

It is clear that the violation of deeply held beliefs and values is a significant consequence of addiction hence, restoring these beliefs and values becomes an important component of recovery (Horvath, A.T, ABPP, Misra, K., Epner, A. K., & Cooper, G.M.; ed. Zupanick, C. E., 2015). It is also clear that a disconnection from the spiritual self or a higher power can lead to addiction just as addiction greatly impact negatively on the same connection to self, others, and higher power.

CHAPTER THREE

TOWARDS A HOLISTIC UNDERSTANDING AND TREATMENT OF DRUG ADDICTION

More and more, the different addiction professionals as well as recovery or drug rehabilitation centers talk about holistic understanding and treatment which is characterized by understanding of the whole human person in terms of body, mind, and spirit or body, mind and soul, in order to treat or heal all aspects of the person's life. Human beings are complex beings with varied personal individualism. This complex nature is manifested in his intake of internal and external impulses as well as in his behavior or reaction to the same impulses. Hence, the growing understanding that health and disease are determined by complex interactions between biological, psychological and sociological factors (Leigh and Reiser, 1980) without overlooking the spiritual factor.

Addiction is not only the disease of the brain, or the disease which affects a whole human person, but also a family and societal disease. Understanding why one becomes addicted to alcohol and other drugs will require how genetics and environment affect the addict; and treatment of such person will put into consideration the genetic and environmental influences which are expressed here as the biopsychosocial-spiritual factors. This will become a holistic approach to treatment and recovery. For instance, *The National Institute on Drug Abuse* (NIDA) holistic drug recovery programs can help patients:

- Identify what is triggering their addictions
- Understand the step-by-step events that led to drug use
- Stop the addiction early
- Cope with triggers through visualization, thought disruption and relaxation
- Find alternatives to drug abuse
- Develop a long-term recovery plan

(Recovery.org, 2015)

Daniel P. Sulmasy, OFM, MD, PhD (2002), in his article *A Biopsychosocial-Spiritual Model for the Care of Patients at the End of Life* stressed that healing a whole person is healing the relationship with self, God, others and environment. He puts it this way:

> On this model (Biopsychosocial-spiritual model), healing is not, as it is often characterized, a "making whole." Rather, healing, in its most basic sense, means the restoration of right relationships. What genuinely holistic health care means then is a system of health care that attends to all of the disturbed relationships of the ill person as a whole, restoring those that can be restored, even if the person is not thereby completely restored to perfect wholeness. A holistic approach to healing means that the correction of the physiological disturbances and the restoration of the milieu interior is only the beginning of the task. Holistic healing requires attention to the psychological, social, and spiritual disturbances as well. As Teilhard de Chardin (1960) puts it, besides the *milieu interior*, there is also a *milieu divin*.

BIOPSYCHOSOCIAL-SPIRITUAL ASSESSMENT

In order to understanding and embrace a holistic approach, the extent and nature of the client's alcohol or substance use as well as its interaction with other life areas, it is important to carefully carry out diagnosis, for appropriate case management, and successful treatment (SAMHSA, 2009). This understanding begins during the screening and assessment process, which helps match the client with appropriate treatment services. There are hundreds of screening instruments and assessment tools which are specifically helpful to case manager or counselors to determine "whether further assessment is warranted, the nature and extent of a client's substance use disorder, whether a client has a mental disorder, what types of traumatic experiences a client has had and what the consequences are, and treatment-related factors that impact the client's response to interventions." (SAMHSA, 2009)

Screening:

The American Society for Addiction Medicine (ASAM) states,

> Screening for alcohol and/or drug misuse is critical to the prevention of or early intervention in addiction. For those at risk of developing a serious problem with drinking or drugs, the identification of early warning signs can be enough to change negative drinking or drug use habits. For others, these assessments are important first steps toward treatment of and recovery from addiction.

The Substance Abuse and Mental Health Services Administration (SAMHSA, 2009) defines *Screening* as "a process for evaluating the possible presence of a particular problem. The outcome is normally a simple yes or no". Screening and intake is the initial contact between a client and the treatment system where the client forms his/her first impression and what to expect in treatment; and how screening is conducted can be as important as the actual information gathered. This "sets the tone of treatment and begins the relationship with the client." (SAMHSA, 2009). It is no doubt that a screening can reveal an outline of a client's involvement with alcohol, drugs, or both, but it neither results in a diagnosis nor provide details of how substances have affected the whole person of the client (SAMHSA, 2009).

It is important that the screening should touch the biopsychosocial spiritual (BPSS) aspect of the client. Hence, according to SAMHSA, the domains to screen for when working with the client include:

- Substance abuse

- Pregnancy considerations

- Immediate risks related to serious intoxication or withdrawal

- Immediate risks for self-harm, suicide, and violence

- Past and present mental disorders, including posttraumatic stress disorder (PTSD) and other anxiety disorders, mood disorders, and eating disorders

- Past and present history of violence and trauma, including sexual victimization and interpersonal violence

- Health screenings, including HIV/AIDS, hepatitis, tuberculosis, and STDs

Other area to screen for include:

- Family History and Relationships
- Spirituality and Religion in terms of sense of meaning and purpose in life.

Assessment:

The Substance Abuse and Mental Health Services Administration (SAMHSA, 2009) defines *Assessment* as "a process for defining the nature of that problem, determining a diagnosis, and developing specific treatment recommendations for addressing the problem or diagnosis."

Assessment should be relational between the counselors or case manager and the client in such a way that therapeutic communication is involved, yet with the client as the main focus since "by enhancing communication and provider-patient relationship, patient-centered interviewing produces the relevant biopsychosocial reality of each patient at each visit" (Smith, R.C., 2002), and the approach puts the client's needs foremost, for example, the client's interests, concerns, questions, ideas, requests (Smith, R.C., 2002).

It is noteworthy however, that "the focus of the assessment may vary depending on the program and the specific issues of an individual client. A structured biopsychosocial history interview can be obtained by using The Psychosocial History (PSH) assessment tool (Comfort et al. 1996), a comprehensive multidisciplinary interview incorporating modifications of the Addiction Severity Index (ASI) designed to assess the history and needs … (of clients) in substance abuse treatment" (SAMHSA, 2009). Efforts have also been made to "retain the fundamental structure of ASI while expanding it to include family history and relationships, relationships with partners, responsibilities for children, pregnancy history, history of violence and victimization, legal issues, and housing arrangements" (Comfort and Kaltenbach 1996, SAMHSA, 2009).

SAMHSA (2009) also noted that the different treatment programs have their own prescribed format for obtaining a psychosocial history that coincides with State regulations as well as other standards set by Joint Commission on Accreditation of Healthcare Organizations (JCAHO) and Commission on Accreditation of Rehabilitation Facilities (CARF). Applying the biopsychosocial and spiritual issues that are pertinent to women as a guide to the biopsychosocial spiritual assessment of alcohol and substance abusers, the following areas are important as presented in SAMHSA's Treatment Improvement Protocol (TIP) Series, No. 51:

- *Medical History and Physical Health*: Review HIV/AIDS status, history of hepatitis or other infectious diseases, and HIV/AIDS risk behavior; explore history of gynecological problems, use of birth control and hormone replacement therapy, and the relationship between gynecological problems and substance abuse; obtain history of pregnancies, miscarriages, abortions, and history of substance abuse during pregnancy; assess need for prenatal care.

- *Substance Abuse History*: Identify people who initially introduced alcohol and drugs; explore reasons for initiation of use and continued use;

discuss family of origin history of substance abuse, history of use in previous and present significant relationships, and history of use with family members or significant others.

- *Mental Health and Treatment History*: Explore prior treatment history and relationships with prior treatment providers and consequences, if any, for engaging in prior treatment; review history of prior traumatic events, mood or anxiety disorders (including PTSD), as well as eating disorders; evaluate safety issues including parasuicidal behaviors, previous or current threats, history of interpersonal violence or sexual abuse, and overall feeling of safety; review family history of mental illness; and discuss evidence and history of personal strengths and coping strategies and styles.

- *Interpersonal and Family History*: Obtain history of substance abuse in current relationship, explore acceptance of client's substance abuse problem among family and significant relationships, discuss concerns regarding child care needs, and discuss the types of support that she has received from her family and/or significant other for entering treatment and abstaining from substances.

- *Family, Parenting, and Caregiver History*: Discuss the various caregiver roles she may play, review parenting history and current living circumstances.

- *Children's Developmental and Educational History* (applicable to women and children programs): Assess child safety issues; explore developmental, emotional, and medical needs of children.

- *Sociocultural History*: Evaluate client's social support system, including the level of acceptance of her recovery; discuss level of social isolation prior to treatment; discuss the role of her cultural beliefs pertaining to her substance use and recovery process; explore the specific cultural attitudes toward women and substance abuse; review current spiritual practices (if any); discuss current acculturation conflicts and stressors; and explore need or preference for bilingual or monolingual non-English services.

- *Vocational, Educational, and Military History*: If employed, discuss the level of support that the client is receiving from her employer; review military history, then expand questions to include history of traumatic events and violence during employment and history of substance abuse in the military; assess financial self-reliance.

- *Legal History*: Discuss history of custody and current involvement with child protective services, if any; obtain a history of restraining orders, arrests, or periods of incarceration, if any; determine history of child placement with women who acknowledge past or current incarceration

Although the above is designed for women, it can be summarized as a Biopsychosocial Spiritual Assessment addressing the following four areas as they contribute to the client's present functioning (as adapted from http://medicine.nevada.edu/Documents/unsom/psy-reno/clerkship/Biopsychosocialspiritual.pdf).

Biological:

- Consideration family history as regards blood relatives with history of Alcohol and other Drugs (AOD).
- Consideration if there was in-utero exposure to AOD or other medications. Also birth complications, such as prematurity, birth trauma or extended period of hospitalization.

- Consideration of medical illnesses or diseases such as Hepatitis C, HIV, Syphilis, Gonorrhea or any sexually transmitted diseases

Psychological:

- Consideration of any form of mental disorders or Co-occurring disorders
- Consideration of the substance use and psychiatric problems impacted the client's development.
- Considerations of previous recovery, relapse or treatment
- Considerations of client's quality of relationships with important figures, such as parents, grandparents, friends, significant teachers or employer
- Considerations of current developmental demands on the client, such as marriage, divorce, children leaving home, loss, aging etc.

Social:

- Consideration of the client's current support system
- Consideration of the cultural influence and peer influence

- Consideration of the client's current status of relationship with important figure or family
- Consideration of the client's housing status and arrangement
- Consideration of legal status and the role of agencies on the client - such as Veteran's Administration, Child Protective Services, and Criminal Justice System

Spiritual:

- Consideration of the role spirituality plays in the life of the client
- Consideration of the client's affiliation with a spiritual or religious community
- Consideration of how spirituality contribute to the client's ability to hope and find meaning in life.

The Indian Health Service (IHS), an operating division within the U.S. Department of Health and Human Services, which is responsible for providing medical and public health services to members of federally recognized Tribes and Alaska Natives developed a comprehensive biopsychosocial assessment named ADULT STANDARD BIOPSYCHOSOCIAL ASSESSMENT (Revised 5/3/2006) which assesses the Biological, Psychological, Social, and

Spiritual functioning of the client with addiction. The full assessment form is found in the APPENDIX B

Biopsychosocial-Spiritual Integration in Treatment

Screening and Assessment help the case manager or the substance abuse counselor to be able to identify the problem areas of the client/patient and as such able to make referral where needed as well as to discover where the focus of the treatment should be. According to the National Institute on Drug Addiction (2012),

> An individual's treatment and services plan must be assessed continually and modified as necessary to ensure that it meets his or her changing needs. A patient may require varying combinations of services and treatment components during the course of treatment and recovery. In addition to counseling or psychotherapy, a patient may require medication, medical services, family therapy (spiritual counseling), parenting instruction, vocational rehabilitation, and/or social and legal services. For many patients, a continuing care approach provides the best results, with the treatment intensity varying according to a person's changing needs

By implication, aiming at achieving effective treatment requires attention to the multiple needs of the addict and not just focusing on the drug addiction treatment. In other words, effective treatment should be biopsychosocial spiritual focused, addressing individual's drug addiction as well as the associated biological, psychological, social, spiritual, vocational, and legal problems; and the treatment should be patient/client's age, gender, ethnicity, and culture appropriated. (NIDA, December 2012)

It is also very important to be aware of the fact that treatment varies depending on the type of drug and the characteristics of the patients, therefore, matching treatment settings, interventions, and services to an individual's particular problems and needs is very critical to his or her eventual success in returning to productive functioning in the family, workplace, and society. (NIDA, December 2012)

Using a biopsychosocial-spiritual approach to treatment, patients in addiction and co-occurring disorder program need to partake in the following treating modalities:

Biological

The biological aspect of treatment of addiction includes detoxification, medication assisted treatment, medical checkups or psychiatric attention if client with co-occurring disorders, and treatment of any addiction related diseases or illnesses.

The patients who are under the influence of alcohol and/or other drugs especially those who exhibiting withdrawal symptoms would need detoxification. However, "medically assisted detoxification is only the first stage of addiction treatment and by itself does little to change long-term drug abuse" (NIDA, December 2012). Hence, patients need to be encouraged to continue drug treatment following detoxification while motivational enhancement and incentive strategies which begun at initial patient intake continues to improve treatment engagement (NIDA, December 2012).

Patients/clients who are Addicted to alcohol or illegal substances and especially those with Co-Occurring Disorders will need to meet with either a psychiatrist or qualified physician at a minimum of once per week so that any medications that have been prescribed can be monitored for therapeutic effectiveness and any necessary changes can be made (http://sierratucson.crchealth.com/addiction/).

Medications remain an important element of treatment for many patients, although with ongoing counseling and other behavioral therapies. Medications such as methadone, buprenorphine, and naltrexone are effective in treating heroin or opioids addicts as they stabilize their lives and reduce their illicit drug use just as acamprosate, disulfiram, and naltrexone are approved medications for treating alcohol dependence. For nicotine addicts, a nicotine replacement product such as patches, gum, lozenges, or nasal spray or an oral medication like bupropion or varenicline can be effective component of treatment when part of a comprehensive behavioral treatment program (NIDA, December 2012).

Effective drug abuse treatment should also addresses some of the drug-related behaviors that put people at risk of infectious diseases such as HIV, Hepatitis B, Hepatitis C and gonorrhea. Take for instance, "substance abuse treatment facilities should provide onsite, rapid HIV testing rather than referrals to offsite testing—research shows that doing so increases the likelihood that patients will be tested and receive their test results. Treatment providers should also inform patients that highly active antiretroviral therapy (HAART) has proven effective in combating HIV, including among drug-abusing populations, and help link them to HIV treatment if they test positive" (NIDA, December 2012).

Psychological

The patients/clients need to be counselled from time to time and to engage in therapeutic activities. Some of them have experienced trauma at one time or the other. According to Michele Rosenthal, (March 30, 2015), "While experiencing a trauma doesn't guarantee that a person will develop an addiction, research clearly suggests that trauma is a major underlying source of addiction behavior". And according to a report issued by the National Center for Post-Traumatic Stress Disorder and the Department of Veterans Affairs, there is a strong correlation between trauma and alcohol addiction (Rosenthal, M., 2015):

- Sources estimate that 25 and 75 percent of people who survive abuse and/or violent trauma develop issues related to alcohol abuse.

- Accidents, illness or natural disasters translate to between 10 to 33 percent of survivors reporting alcohol abuse.

- A diagnosis of PTSD (post-traumatic stress disorder) increases the risk of developing alcohol abuse.

- Female trauma survivors who do not struggle with PTSD face increased risk for an alcohol use disorder.

- Male and female sexual abuse survivors experience a higher rate of alcohol and drug use disorders compared to those who have not survived such abuse.

Hence, trauma counseling and exercises, behavioral therapies, including individual, family, or group counseling—are the most commonly used forms of drug abuse treatment (NIDA, December 2012). These may involve healing the patient's trauma, addressing a patient's motivation to change, providing incentives for abstinence, building skills to resist drug use, replacing drug-using activities with constructive and rewarding activities, improving problem-solving skills, and facilitating better interpersonal relationships with spouse, children and other family members, as well as actively participating in group therapy and other peer support programs during and following treatment.

Social

Substance use usually impacts multiple aspects of an addict's life as such patients/clients often enter treatment with multiple problems such as homelessness, unemployment, and broken family or social relationships (SAMHSA, September 11, 2014). Hence, the need for supportive services become critical components of a behavioral health system which can help patients meet their treatment goals (SAMHSA, October 16, 2014). By this, the patients are coordinated and supported in resolving their housing, employment, education, and other supports they may need. "Frequently, when individuals are involved in multiple public systems it is important for a single point of contact to coordinate care and engage all the system partners in service planning and delivery. For young people, this is often done through a wraparound process" (SAMHSA, October 16, 2014).

Treatment also involve the patients participating in meaningful daily activities, such as a job, school, volunteerism, family caretaking, or creative endeavors, and having the independence, income, and resources to participate in society; and engaging in relationships and social networks that provide support, friendship, love, and hope (SAMHSA, September 11, 2014). "One way that treatment programs facilitate therapeutic networking, foster healthy relationships, and provide targeted services is to offer specially designed substance abuse treatment programs or groups for specified types of clients. In addition, *Community*-related services also included counseling, ancillary, and pre-treatment services" (SAMHSA, September 11, 2014).

In 2012, nearly three-quarters (74 percent) of facilities provided services in support of social skills development. Employment counseling or training for clients was offered by 37 percent of facilities, and only 7 percent provided child care services in the United States (SAMHSA, September 11, 2014).

Spiritual

Psychologists are now developing and evaluating a variety of spiritually integrated approaches to treatment, which includes but not limited to forgiveness programs to help people come to terms with bitterness and anger;

programs to help survivors of sexual abuse deal with their spiritual struggles; treatments for people with eating disorders focusing on their spiritual resources; and programs that help alcoholics or drug abusers re-connect to their higher power (Pargament, K. I., 2013).

In the same vein, the World Health Organization has declared that spirituality is an important dimension of quality of life (WHOQOL Grooup, 1995; Sulmasy, D. P., 2002) for how one is doing spiritually affect's ones physical, psychological, and social status (Sulmasy, D. P., 2002). And of course, effective care depends also on basic knowledge about religious and spiritual diversity, understanding of how religion and spirituality are interwoven into adaptive and maladaptive human behavior, and skills in assessing and addressing religious and spiritual issues that arise in treatment (Pargament – K. I, March 22, 2013)

Many scientific studies have come to support the importance of spirituality in recovery programs that according to the research by Project MATCH, sponsored by the National Institute on Alcohol Abuse and Alcoholism, "spirituality focused addiction treatment programs have resulted in up to a 10 percent greater abstinence rates than other forms of treatment. Other studies indicate an inverse relationship between religious involvement and substance dependence, as well as an inverse relationship between meditation practices and substance use" (Alternatives in Treatment, 2014)

To this effect, it is important to explore the spiritual heritage and belief of the patient and how important is spirituality to him or her. This is because spiritual concerns are important to many clients (Sulmasy, D.P., 2002), and majority of people have a unique personal understanding about the meaning and purpose of life (Horvath, A.T, ABPP, Misra, K., Epner, A. K., & Cooper, .M.; ed. Zupanick, C. E., 2015). Spiritual practices such as prayer, contemplation, yoga, Zen, and transcendental meditation impact physiological processes in the brain (van Wormer & Davis, 2003).

Although not all the support groups explore spiritual approach, Alcoholics Anonymous was among the first recovery program to explicitly connect spirituality with addiction treatment and recovery; and linking the spiritual elements of addiction recovery with the physical and psychological aspects through the 12-step model (Alternatives in Treatment, 2014). The spiritual model often refers to the use of alcohol or drugs as a way to fill the spiritual emptiness that overwhelmed the addict, hence the reason why many advocate turning to God or a higher power in other to fill this emptiness, find meaning in life and thereby overcome addiction. The Alcoholic Anonymous and the similar inpatient or outpatient programs according to *Alternatives in Treatment* (2014) emphasize:

- Experiencing addiction as a spiritual, mental, physical, and social symptom
- Believing and trusting in a power that's greater than one's own willpower
- Exploring and re-evaluating one's purpose in life
- Committing to moral and ethical behavior
- Developing hope that one can recover from addiction
- Performing an honest, thorough moral inventory of oneself
- Admitting wrongs to oneself, others, and a higher power
- Making amends to those that one has hurt in the past
- Using prayer, meditation to consciously connect with a higher power

Taking into the consideration the fact that addiction affect all aspect of human person and human interaction, the biological, psychological, social, and spiritual interventions or treatments should not be taking as separate approaches, rather as integrated method of treatment where the patients/clients are treated or help into recovery touching all the aspects that may apply according to the individual client's need. Meaning that the biopsychosocial-spiritual model is an integrated method of treatment which takes into account the diversity of client biological, psychological, social and spiritual needs. The biopsychosocial-spiritual factors of treatment are considered concurrently during the treatment.

CONCLUSION

Holding to the fact that environmental factors interact with the person's biology and affect the extent to which genetic factors exert their influence, the resiliencies acquired by individuals through parenting or later life experiences can affect the extent to which genetic predispositions lead to the behavioral and other manifestations of addiction; and considering the role culture (socio-religious environment) plays in addiction becomes actualized in persons with biological vulnerabilities to the development of addiction" (ASAM, 2011). Hence, the question "why people get addicted to alcohol and drugs?" can only be substantially answered based on the understanding of the biological, psychological, social, and spiritual factors of addiction.

Different factors are responsible for alcohol and drug addictions and as there are many addicts there are many factors which cause their addiction. And in many occasions, there is not one cause of addiction, it is most caused by interaction of and interwoven biopsychosocial-spiritual factors and the effects of same addiction are biopsychosocial-spiritual in nature. Hence, to effectively treat addiction, one needs to understand what part biological, psychological, social and spiritual factors played in the addiction of the patients and as such one has to take all those factors into consideration during treatment – not fragmentally rather, in a well-integrated manner where all the aspects of biopsychosocial-spiritual treatments are carried out concurrently. In this way, the holistic treatment of an addict may be met.

How exactly?

Back to Adeolu's case study, case management and different services coordination need to be implemented to provide comprehensive and integrated or concurrent treatments which may include appropriate referrals as needed. A collaborative treatment coordinated by the substance abuse program case management with the medical team, psychiatrist, psychotherapists, and necessary legal, social spiritual/religious services will be the key to integrated biopsychosocial-spiritual treatment. Hence the treatment plan will include but not limited to:

- Adeolu will continue to remain clean and sober
- Adeolu will continue to see his psychiatrist for his bipolar disorders and depression, and take his medications
- Adeolu will continue to see his primary care physician for his HIV and liver cirrhosis treatment, and be faithful to his medications
- Adelolu will be assisted to reconnect with his children
- Adeolu will be supported in his housing problem
- Adeolu will attend and participate in AA and/or NA twice a week
- Adeolu will be referred to his spiritual/religious tradition where he may be helped to reconnect with God
- Adeolu will be assisted with job skills, and career opportunities

While the focus of the treatment is on recovery, every other aspect of the client's life which is affected by the client's addiction is also addressed.

REFERENCES

Alternative in Treatment (Feb. 26, 2014). Spirituality & addiction recovery. Retrieved 8/10/2015

    from http://www.alternativesintreatment.com/blog-page-posts/spirituality-addiction-recovery/

Amodia, D.S., Cano, C., & Eliason, M.J. (2005). An integral approach to substance abuse. The Journal of Psychoactive Drugs, Vol. 37 (4), 363-371.

ASAM, (2011) Definition of Addiction. Retrieved July 16, 2015

    from http://www.asam.org/for-the-public/definition-of-addiction

"   Screening and Assessment. Retrieved 8/10/15

    from http://www.asam.org/for-the-public/screening-and-assessment

Atkins, Jr. R.G. & Hawdon, J.E. (2008). Religiosity and participation in mutual-aid support groups for addiction. Journal of Substance Abuse Treatment, 33(3), 321-331 doi: 10.1016/i.isat.2007.07.001

Beck, T., (Tuesday, Nov. 5, 2013). Genetics of addiction: Twin studies. *VCU News*. Retrieved 7/27/15 from http://news.vcu.edu/article/Genetics_of_addiction_Twin_studies

Bevilacqua L. & Godlman, D., (2009). Genes and

    Addictions. *National Center for Biotechnology Information (ncbi)*. Retrieved 7/27/15 from

CASAColumbia, (2005). Family matters: Substance abuse

   and the American family. Retrieved 8/4/15 from
   http://www.casacolumbia.org/addiction-
   research/reports/family-matters-substance-abuse-
   and-american-family

"   (2012). Addiction as a disease. Addiction
   medicine: Closing the gap between
   science and practice. Retrieved 7/27/15 from
   http://www.casacolumbia.org/addiction/disease-
   model-addiction
Center for Disease Control and Prevention (CDC). *Alcohol-*
   Related Disease Impact (ARDI). Atlanta, GA:CDC
Diagnostic and Statistical Manual of Mental Disorder-Fifth
   Edition. (2013).Washington, DC: American
   Psychiatric Publishing.
First Step Recovery Homes, Inc., (2012). Who we are.
   Retrieved 8/14/15 from
   http://www.firststeprecoveryhomes.org/about_us
Galanter, M. (2006). Spirituality and addiction: A research
   and clinical perspective. American
   Academy of Addiction Psychiatry, 15: 286-292.
   doi: 10.1080/10550490600754325
Garfolo, B., (2015). Merging spirituality and clinical
   psychology: The biopsychosocial-spiritual
   model of mental health. Retrieved 7/20/2015 from
   https://www.academia.edu/4814375/Merging_Spirit
   uality_and_Clinical_Psychology_The_BioPsychoSo
   cial-Spiritual_Model_of_Mental_Health
Gerrig, Richard J. & Philip G. Zimbardo (2002). Glossary
   of psychological terms. *American*
   Psychological Association. Retrieved 7/31/2015
   from
   http://www.apa.org/research/action/glossary.aspx?ta
   b=16
Griffiths, M. (2005). A 'components' model of addiction

within a biopsychosocial framework.
The Journal of Substance Use, 10(4), 191-197. doi: 10.108014659890500114359

Hatala, A.R. (2012). The status of the 'biopsychosocial' model in health psychology: Towards an integrated approach and a critique of cultural conceptions. The Journal of Medicine Psychology, 1, 51-62. doi: org/10.4236/ojmp.2012.14009

Horvath, A.T, ABPP, Misra, K., Epner, A. K., & Cooper, G.M., (2015). Zupanick, C. E. (ed).
The Spirituality of addiction and recovery.
Retrieved 7/16/2015 from
http://www.amhc.org/1408-addictions/article/48424-the-spirituality-of-addiction-recovery

"   The Spirituality of addiction and recovery continued. Retrieved 7/20/2015 from
http://www.amhc.org/1408-addictions/article/48426-the-spirituality-of-addiction-recovery-continued

"   Incorporating spirituality into recovery from addiction. Retrieved 7/20/2015

from http://www.amhc.org/1408-addictions/article/48425-incorporating-spirituality-into-recovery-from-addiction

"   Disease model of addiction and recovery implications. Retrieved 7/27/15

from http://www.amhc.org/1408-addictions/article/48343-disease-model-of-addiction-and-recovery-implications

"   Psychological Causes of Addiction. Retrieved 7/31/15 from http://www.amhc.org/1408-addictions/article/48345-psychological-causes-of-addiction

"   (August 26, 2013). Addiction: Sociological causes of addiction and temperance model. *Centersite.net.* Retrieved 8/3/15 from http://www.centersite.net/poc/view_doc.php?type=doc&id=48351&cn=1408

IHS (Revised 5/3/2006). Adult Standard Biopsychosocial Assessment. Retrieved 8/12/15 from ftp://ftp.ihs.gov/pubs/ehr/Templates/TIU%20Note%20Templates/Nationally%20Approved%20EHRTemplates/Behavioral%20Health/ADULT%20STANDARD%20BIOPSYCHOSOCIAL%20Template.pdf

Karl Marx (1976), *Theses on Feuerbach* in: K. Marx and F. Engels, *Collected Works*, Vol. 5, Progress Publishers, Moscow, p. 4.

Kumpfer, K. L., Trunnell, E. P., & Whiteside, H. O., (2015). Engs. R., (ed). The biopsychosocial model: application to the addiction field. Retrived 7/20/2015 from http://www.indiana.edu/~engs/cbook/chap7.html http://www.ncbi.nlm.nih.gov/pmc/articles/PMC2715956/

Leigh, H. and Reiser, M.F. (1980) *Approach to patients: The systems contextual framework and the Patient evaluation grid. The Patient: Biological, Psychological, and Social Dimension of Medical Practice.* Plenum Medical Book Company: New York, 185Life Progress. *The meaning of addiction.* Retrieved July 16, 2015 from http://lifeprocessprogram.com/the-meaning-of-addiction-1-the-concept-of-addiction/

Marlow K., (April 10, 2013). OCD, Tourette's syndrome and Addiction: A Real Distinction?

Psychology Today. Retrieved 8/16/2015 from
https://www.psychologytoday.com/blog/the-
superhuman-mind/201304/ocd-tourettes-syndrome-
and-addiction-real-distinction

Miller, W. (1998). Motivational interviewing: Towards a
motivational definition and
understanding of addiction. Newsletter for Trainers,
vol 5, No. 3 p. 2-6. Retrieved May 10, 2016, from:
http://motivationalinterview.net/clinical/motmodel.h
tml

Miller, W.R., Forcehimes, A., O'Leary, M., &LaNoue,
M.D. (2008). Spiritual direction in
addiction treatment: Two clinical trials. The Journal
of Substance Abuse Treatment, 35(4), 434-442. doi:
10.1016/i.isat.2008.02.004

Mooney, M., (February 26, 2012). The spirituality of
addiction. *MBG.* Retrieved 8/7/2015
from http://www.mindbodygreen.com/0-4094/The-
Spirituality-of-Addiction.html

National Council on Alcoholism and Drug Dependence,
Incorporation (NCADD). Alcohol and
Drug Abuse Affect Everyone in the Family.
Retrieved 8/4/15 from https://ncadd.org/get-
help/family-information-and-education/144-family-
education

National Institute on Drug Abuse (NIDA). *The Science of
Drug Abuse and Addiction: The Basics.* Retrieved
7/20/15 from
http://www.drugabuse.gov/publications/media-
guide/science-drug-abuse-addiction-basics.

"   (July 2014). Drugs, Brains, and Behavior: The
Science of Addiction. Retrieved 7/22/2015 from
http://www.drugabuse.gov/publications/drugs-
brains-behavior-science-addiction/drug-abuse-
addiction

"   (July 29, 2015). NIDA and NIAAA commentary strongly supports brain disease model of addiction. Retrieved 8/11/2015 from http://www.niaaa.nih.gov/news-events/news-noteworthy/nida-and-niaaa-commentary-strongly-supports-brain-disease-model

"   (December, 2012)   Cognitive-Behavioral Therapy (Alcohol, Marijuana, Cocaine, Methamphetamine, Nicotine) Retrieved 7/31/15 from http://www.drugabuse.gov/publications/principles-drug-addiction-treatment-research-based-guide-third-edition/evidence-based-approaches-to-drug-addiction-treatment/behavioral

"   (December 2012). Principles of Drug Addiction Treatment: A Research-Based Guide (Third Edition). Retrieved 8/13/2015 from http://www.drugabuse.gov/publications/principles-drug-addiction-treatment-research-based-guide-third-edition/principles-effective-treatment

"   Quick Screening Question. Retrieved 8/12/15 from http://www.drugabuse.gov/sites/default/files/files/QuickScreen_Updated_2013%281%29.pdf

National Drug Intelligence Center, (2011). *The Economic Impact of Illicit Drug Use on American Society.* Washington, DC: United States Department of Justice.

Ogborne, Alan C. (2015). Theories of Addiction and Implications for Counselling. Retrieved 7/20/15 from http://knowledgex.camh.net/amhspecialists/guidelines_materials/adp/Documents/adp_chapter1.pdf

Pargament – K. I, (March 22, 2013). What role do religion and spirituality play in mental health? *American Psychological Association.* Retrieved 8/7/2015 from http://www.apa.org/news/press/releases/2013/03/religion-spirituality.aspx

Perkins, C., (2015). Addiction and Spirituality. Retrieved 8/7/2015 from http://www.alternatives-for-alcoholism.com/addiction-and-spirituality.html

Robbins, T.W. & Everitt, B.J. (1999). Drug addiction: Bad habits add up. News and View Feature, vol. 398. Retrieved from: www.nature.com

Rogers Memorial Hospital, (February 24, 2014). DSM-5 Now Categorizes Substance Use Disorders in a Single Continuum. Retrieved 7/20/15 from https://rogershospital.org/blog/dsm-5-now-categorizes-substance-use-disorders-single-continuum

Rosenthal, M, (March 30, 2015). Trauma and addiction: 7 reasons your habit makes perfect sense. *Recovery.Org.* Retrieved 8/13/15 from http://www.recovery.org/pro/articles/trauma-and-addiction-7-reasons-your-habit-makes-perfect-sense/

SAHMSA, (2013). Results from the 2013 National Survey
      on Drug Use and Health: Summary
      of National Findings. Retrieved 8/16/2015 from
      http://www.samhsa.gov/data/sites/default/files/NSD
      UHresultsPDFWHTML2013/Web/NSDUHresults2
      013.pdf

"   (2009). Substance Abuse and Mental Health
      Services Administration (US)
      Treatment Improvement Protocol (TIP) Series, No.
      51. Retrieved 8/12/15 from
      http://www.ncbi.nlm.nih.gov/books/NBK83253/

"   (2009). *The Evidence: Integrated Treatment for Co-*
      Occurring Disoders. DHHS Publication
      No. SMA-08-4366. Retrieved 8/16/2015 from
      https://store.samhsa.gov/shin/content/SMA08-
      4367/TheEvidence-ITC.pdf

"   (September 11, 2014). Recovery Services Provided
      by Substance Abuse Treatment
      Facilities in the United States *The N-SSAT Report.*
      Retrieved 8/13/15 from
      http://www.samhsa.gov/data/sites/default/files/NSS
      ATS-SR175-RecoverySvcs-2014/NSSATS-SR175-
      RecoverySvcs-2014.htm

"   (Updated October 16, 2014). Behavioral Health
      Treatment and Services. Retrieved
      8/13/15 from http://www.samhsa.gov/treatment

Sierra Tucson. Substance Abuse and Treatment Center.
      Retrieved 8/13/15 from
      http://sierratucson.crchealth.com/addiction/

Sinha, R. (2011). New findings on biological factors
      predicting addiction relapse vulnerability.
      Current Psychuatry Rep. 13(5), 398-405. doi:
      10.1007/s11920-011-0224-0

Smith, R.C., (2002). The Biopsychosocial Revolution. J Gen Intern Med. 2002 Apr; 17(4): 309–310. doi: 10.1046/j.1525-1497.2002.20210.x

Stoddart, T., (May 8, 2013). Addiction: a disease of spirituality. *Sobernation.* Retrieved 8/7/2015 from http://www.sobernation.com/addiction-a-disease-of-spirituality/-200.

Substance Abuse and Mental Health Services Administration. (2012). *Comprehensive case management for substance abuse treatment.* (HHS Publication No. SMA 12–4215, Treatment Improvement Protocol [TIP] Series No. 27). Rockville, MD: Author.

Sulmacy, D. P., (2002). A Biopsychosocial-Spiritual Model for the Care of Patients at the End of Life. *The Gerontologist.* Vol 42, Special Issue Ill, 24-33 Retrieved 8/12/15 from http://pmr.uchicago.edu/sites/pmr.uchicago.edu/file s/uploads/Sulmasy_ABiopsychosocial-SpiritualModelfortheCareofPatientsattheEndofLife %20.pdf

Sunshine Coast Health Center. Image of Biopsychosocial Spiritual Treatment – Retrieved 8/12/2015 from http://www.sunshinecoasthealthcentre.ca/addiction-treatment/

The CRAFFT Screening interview. Retrieved 8/12/2015 from http://www.ceasarboston.org/CRAFFT/pdf/CRAFF T_English.pdf

van Wormer, K & Davis, D. (2003). Addiction treatment: A strengths perspective. Pacific Grove, CA: Brooks/Cole.

Weinberg, D. (2011). Sociological perspectives on addiction. *Academia.edu.* Retrieved 8/3/15

from
https://www.academia.edu/4433244/Sociological_P
erspectives_on_Addiction

World Health Organization WHO, (2015). The determinant
of health. *Health Impact Assessment
(HIA).* Retrieved 8/14/15 from
http://www.who.int/hia/evidence/doh/en/

"   (2015). Prevalence of drug use disorders. Global
Health Observatory (GHO) data.
Retrieved 8/16/2015 from
http://www.who.int/gho/substance_abuse/burden/dr
ug_prevalence_text/en/

Zickler, P., (November, 1999). Twin studies help define the
role of genes in vulnerability to drug
abuse. *NIDA Notes.* Vol. 14, No. 4. Retrieved
7/27/15 from
http://archives.drugabuse.gov/NIDA_Notes/NNVol
14N4/Twins.html

APPENDIX A

BIOPSYCHOSOCIAL HISTORY (SCREENING)
(http://www.signpostrelationshipsolutions.com/resources/SPRSI+Biopsych osocial+history.pdf)

BIOPSYCHOSOCIAL HISTORY

Name_____ Age_____   Date_____

Other Family Members:

Name_____ Age_____

Name_____ Age_____

Name_____ Age_____

Name_____ Age_____

Full Address_____

_____

Home Phone_____ Work_____ E-mail_____

Physical History (please be accurate, medical records may need to be disclosed at some point)

General Health_____

Are you now under a doctor's care?_____If yes, name of doctor_____

Reason for doctor's care_____

Are you taking any medication?_____If yes, what kind?_____

Reason for medication_____Last medical examination _____

Have you ever been hospitalized for a physical illness?_____Describe_____

_____

Have you ever been hospitalized for a mental illness?_____Describe_____

_____

Any recent major illnesses or surgeries?_____

Any recurrent or chronic conditions?_____

Do you smoke _____Do you take drugs?_____If yes, what kind?_____

Do you drink?_____How much?_____

Any Previous Therapy/Counseling?_____If yes, describe, when, where, how long, what for_____

_____

What do you hope to achieve with therapy?_____

_____

Work History

Occupation_____ How long?_____

If presently unemployed, describe the situation_____

_____

Hobbies/Avocations_____

Family Systems Information

Where born_____How long there_____Ethnic ID_____

Parents: Father alive_____Where residing_____Relationship_____

Mother alive_____Where residing_____Relationship_____

Marital Status_____#of marriages_____Spouse's name_____

Living with a partner_____How long_____Partner's Name_____

Children:#1 M F Age_____ #2 M F Age_____ #3 M F Age_____ #4 M F Age_____ #5 M F Age_____

Siblings: Circle your place in the family. If a sibling is deceased, put an X through the placement number.

#1 M F Age____ #2 M F Age____ #3M F Age____ #4 M F Age____ #5 M F Age____ #6 M F Age_____

Family Alcoholism or Domestic Violence?_____ Sexual Addictions or Abuse?_____

Parents divorced?_____If yes, what year_____Your age at the time_____

If deceased, what year?_____Your age at the time_____Cause of death_____

Any step-parents?_____If yes, describe when and your relationship with them_____

_____

If reared by someone other than your birth parents, describe the situation in some detail_____

_____

Tell anything else in the space below that you think would be helpful for me, as your therapist, to know.

Spiritual History

Religious upbringing_____Present Affiliation_____

Is this an important part of your life_____Why/whynot_____

Emotional Status

Are you currently experiencing strong emotions? _____If yes, describe_____

_____

Do you make decisions based on your emotions?_____How well does that work for you?_____

_____

Did you have what you would consider to be childhood or other traumas?_____If yes, describe_____

_____

Have you been treated for emotional disturbances?_____If yes, when?_____

Have you had any thoughts of suicide_____If so, when_____Do you have any thoughts now_____

Present Situation

Please state why you decided to come for counseling/therapy_____

_____

What is the nature of your situation_____

_____

What would you like to experience that is different from what you are experiencing now_____

_____

How long has this been a problem for you_____

_____

Please state what you would like to work on in therapy_____

_____

_____

APPENDIX B

BIOPSYCHOSOCIAL-SPIRITUAL ASSESSMENT

Originally ADULT STANDARD BIOPSYCHOSOCIAL ASSESSMENT

ftp://ftp.ihs.gov/pubs/ehr/Templates/TIU%20Note%20Templates/Nationally%20Approved%20EHRT
emplates/Behavioral%20Health/ADULT%20STANDARD%20BIOPSYCHOSOCIAL%20Template.p
df

BIOPSYCHOSOCIAL ASSESSMENT

Demographics

Client Name:	Date:
Current Address: Street City/State Zip Code	Phone #: ( ) -

Date of Birth:	Marital/Relationship Status:
Nation/Tribe/Ethnicity:	
Primary language of client:	Secondary:
Referral Source:	Phone:
Emergency Contact:	Phone:

Family Relationships

Does the client have any children?

Name	Age	Date of Birth	Sex	Custody? Y/N	Lives With?	Additional Information

Who else lives with the client? (Include spouses, partners, siblings, parents, other relatives, friends)

Name	Age	Sex	Relationship	Additional Information

Primary language of household/family:	Secondary:

Family History

Family History of (select all that apply):

	Mother	Father	Siblings	Aunt	Uncle	Grandparents
Alcohol/Substance Abuse	☐	☐	☐	☐	☐	☐
History of Completed Suicide	☐	☐	☐	☐	☐	☐
History of Mental Illness/Problems such as:	☐	☐	☐	☐	☐	☐
Depression	☐	☐	☐	☐	☐	☐
Schizophrenia	☐	☐	☐	☐	☐	☐
Bipolar Disorder	☐	☐	☐	☐	☐	☐
Alzheimer's	☐	☐	☐	☐	☐	☐
Anxiety	☐	☐	☐	☐	☐	☐
Attention Deficit/Hyperactivity	☐	☐	☐	☐	☐	☐
Learning Disorders	☐	☐	☐	☐	☐	☐
School Behavior Problems	☐	☐	☐	☐	☐	☐
Incarceration	☐	☐	☐	☐	☐	☐
Other	☐	☐	☐	☐	☐	☐
Comments:						

Revised 5/3/06

BIOPSYCHOSOCIAL ASSESSMENT

Critical Population (choose all that apply)

Funding Source	Residential	Legal Involvement
☐ Food Stamp Recipient	☐ Homeless	☐ Protective Services (APS/CPS)
☐ TANF Recipient	☐ Shelter Resident	☐ Court Ordered Services
☐ SSI Recipient	☐ Long Term Care Eligibility	☐ On Probation
☐ SSDI Recipient	☐ Long Term Care Resident	☐ On Parole
☐ SSA (retirement) Recipient		☐ On Pre-Release
☐ Other Retirement Income	**Disability**	☐ Mandatory Monitoring
☐ Medicaid Recipient	☐ Physical Disability	
☐ Medicare Recipient	☐ Severely Mentally Ill	**Other**
☐ General Assistance	☐ SED	☐ Currently pregnant
	☐ Developmentally Disabled	☐ Woman w/dependents
	☐ Chronically Mentally Ill	
	☐ Regional Behavioral Health Authority	

Contact Information
(Secure consents for agency contacts, when possible)

Name of Caseworker	Agency	Phone number

Client's/Family's Presentation of the Problem:

Client's/Family's Expected Outcome:

Physical Functioning

Allergies (Medication & Other):

Current Medical Conditions:

Current Medications (include herbs, vitamins, & over-the-counter):

Past Medications:

Past Medical History including hospitalizations/residential treatment (list all prior inpatient or outpatient treatment including RTC, group home, therapeutic foster care, aftercare, inpatient psychiatric, outpatient counseling):

Dates	Inpt/Outpt	Location	Reason	Completed? Y/N

Surgeries:

BIOPSYCHOSOCIAL ASSESSMENT

Pain Questionnaire

Pain Management: Is the client in pain now? ☐ Yes ☐ No
If yes, ask client to rate the pain on a scale of 1-10 (with 10 being the severest) and enter score here

Is the client receiving care for the pain? ☐ Yes ☐ No
If no, would the client like a referral for pain management? ☐ Yes ☐ No

Nutrition

Nutritional Status: Current Weight   Current Height   BMI

Appetite: ☐ Good	☐ Fair	☐ Poor, please explain below
☐ Recently gained/lost significant weight		☐ Binges/overeats to excess
☐ Restricts food/Vomits/over-exercises to avoid weight gain		☐ Special dietary needs
☐ Hiding/hording food		☐ Food allergies

Comments

Social

Supportive Social Network? (Rate the network using a scale of 1 Weak to 5 Strong)

Immediate Family		Extended Family	
Friends		School	
Work		Community	
Religious		Other	

Comment:

Living Situation:

☐ Housing Adequate	☐ Housing Dangerous	☐ Ward of State/Tribal Court	☐ Dependent on Others
☐ Housing Overcrowded	☐ Incarcerated	☐ Homeless	☐ At Risk of Homelessness

Additional Information:

Employment: Currently Employed?

☐ Yes	Employer		Length of Employment	
☐ Satisfied	☐ Dissatisfied	☐ Supervisor Conflict	☐ Co-worker Conflict	
☐ No	Last Employer:		Reason for Leaving:	

☐ Never Employed	☐ Disabled	☐ Student	☐ Unstable Work History

Financial Situation:

Presence or absence of financial difficulties: (Fields below are optional)

☐ No Current Problems	☐ Large Indebtedness	☐ Relationship Conflicts Over Finances
☐ Impulsive Spending	☐ Poverty or Below	☐ Financial Difficulties

Source of Income (choose all that apply)

Employed: ☐ Full-time ☐ Seasonal ☐ Self-Employed	☐ Part-time ☐ Temporary	Unemployed: ☐ Actively seeking work ☐ Not looking for work	☐ Public Assistance
☐ Retirement	☐ SSD	☐ SSDI	☐ SSI
☐ Medical Disability via Employer		☐ Other:	

Military History:

☐ Never enlisted in Armed Forces, OR			
☐ Branch of Service:		Combat: ☐ Yes	☐ No
Type of Discharge: ☐ Honorable	☐ Dishonorable	☐ Medical	☐ Other:

Sexual Orientation:

☐ Heterosexual	☐ Bisexual
☐ Homosexual	☐ Transgendered
☐ N/A at this time	☐ Comment:

BIOPSYCHOSOCIAL ASSESSMENT

Family Social History

Describe family relationships & desire for involvement in the treatment process:

Perceived level of support for treatment? (scale 1-5 with 5 being the most supportive)

Legal Status Screening

Past or current legal problems (select all that apply)?

☐ None	☐ Gangs	☐ DUI/DWI
☐ Arrests	☐ Conviction	☐ Detention
☐ Jail	☐ Probation	☐ Other:

If yes to any of the above, please explain:

Any court-ordered treatment?	☐ Yes (explain below)	☐ No	
Ordered by	Offense		Length of Time

Education

Educational Level (select one): ☐ less than 12 years – enter grade completed	☐ Some college or tech school
☐ Unknown	☐ High School Grad/GED	☐ College Graduate

If still attending, current School/Grade:

Vocational School/Skill Area:

College/Graduate School – Years Completed/Major:

Leisure & Recreation

Which of the following does the client do? (Select all that apply)

Spend Time with Friends	☐	Sports/Exercise	☐
Classes	☐	Dancing	☐
Time with Family	☐	Hobbies	☐
Work Part-Time	☐	Watch Movies/TV	☐
Go "Downtown"	☐	Stay at Home	☐
Listen to Music	☐	Spend Time at Clubs/Bars	☐
Go to Casinos	☐	Other:	☐

What limits the client's leisure/recreational activities?

Functional Assessment

Is client able to care for him/herself? ☐ Yes   ☐ No   If No, please explain:

Uses or Needs assistive or adaptive devices (select all that apply):

☐ None	☐ Glasses	☐ Walker	☐ Braille
☐ Hearing Aids	☐ Cane	☐ Crutches	☐ Wheelchair
☐ Translated Written Information	☐ Translator for Speaking	☐ Other:	

Does the client have a history of falls? ☐ Yes ☐ No   Explain:

Revised 5/3/06

BIOPSYCHOSOCIAL ASSESSMENT

Psychological

History of Depressed Mood:	☐ Yes	☐ No

History of irritability, anger or violence (tantrums, hurts others, cruel to animals, destroys property):

Sleep Pattern: Number of hours per day _____ Time to onset of sleep? _____

☐ Normal	☐ Sleeping too much	☐ Sleeping too little
Ability to Concentrate: ☐ Normal	☐ Difficulty concentrating	
Energy Level: ☐ Low	☐ Average/Normal	☐ High

History of/Current symptoms of PTSD (re-experiencing, avoidance, increased arousal)? Select all that apply

☐ Intrusive memories, thoughts, perceptions	☐ Nightmares	☐ Flashbacks
☐ Avoiding thoughts, feelings, conversations	☐ Numbing/detachment	☐ Restricted display of emotions
☐ Avoiding people, places, activities	☐ Poor sleep	☐ Irritability
☐ Hypervigilance	☐ Other:	

Any additional information:

Bereavement/Loss & Spiritual Awareness

Please list significant losses, deaths, abandonments, traumatic incidents:

Spiritual/Cultural Awareness & Practice

Knowledgeable about traditions, spirituality, or religion?   ☐Yes   ☐No   Comment:

Practices traditions, spirituality, or religion?   ☐Yes   ☐No   Comment:

How does client describe his/her spirituality?

Does client see a traditional healer?   ☐Yes   ☐No   Comment:

BIOPSYCHOSOCIAL ASSESSMENT

Abuse/Neglect/Exploitation Assessment

History of neglect (emotional, nutritional, medical, educational) or exploitation? ☐Yes ☐No
If yes, please explain:

Has client been abused at any time in the past or present by family, significant others, or anyone else?)   ☐ No   ☐ Yes, explain:

Type of Abuse	By Whom	Client's Age(s)	Currently Occurring? Y/N
Verbal Putdowns			
Being threatened			
Made to feel afraid			
Pushed			
Shoved			
Slapped			
Kicked			
Strangled			
Hit			
Forced or coerced into sexual activity			
Other			

Was it reported? ☐Yes ☐No	To whom?
Outcome	

Has client ever witnessed abuse or family violence? ☐ No ☐ Yes, explain:

Revised 5/3/06

BIOPSYCHOSOCIAL ASSESSMENT

Behavioral Assessment

Abuse/Addiction – Chemical & Behavioral				
Drug	Age First Used	Age Heaviest Use	Recent Pattern of Use (frequency & Amount, etc)	Date Last Used
Alcohol				
Cannabis				
Cocaine				
Stimulants (crystal, speed, amphetamines, etc)				
Methamphetamine				
Inhalants (gas, paint, glue, etc)				
Hallucinogens (LSD, PCP, mushrooms, etc)				
Opioids (heroin, narcotics, methadone, etc)				
Sedative/Hypnotics (Valium, Phenobarb, etc)				
Designer Drugs/Other (herbal, Steroids, cough syrup, etc)				
Tobacco (smoke, chew)				
Caffeine				

Ever injected Drugs? ☐ Yes ☐ No **If Yes, Which ones?**

Drug of Choice?

Consequences as a Result of Drug/Alcohol Use (select all that apply)

☐ Hangovers	☐ DTs/Shakes	☐ Blackouts	☐ Binges
☐ Overdoses	☐ Increased Tolerance (need more to get high)	☐ GI Bleeding	☐ Liver Disease
☐ Sleep Problems	☐ Seizures	☐ Relationship Problems	☐ Left School
☐ Lost Job	☐ DUIs	☐ Assaults	☐ Arrests
☐ Incarcerations	☐ Homicide	☐ Other:	

Longest Period of Sobriety? **How long ago?**

Triggers to use (list all that apply):

Has client traded sex for drugs? ☐ No ☐ Yes, explain:

Has client been tested for HIV? ☐ Yes ☐ No

If yes, date of last test: **Results:**

Has client had any of the following problem gambling behaviors? Select all that apply:

☐ Gambled longer than planned	☐ Gambled until last dollar was gone
☐ Lost sleep thinking of gambling	☐ Used income or savings to gamble while letting bills go unpaid
☐ Borrowed money to gamble	☐ Made repeated, unsuccessful attempts to stop gambling
☐ Been remorseful after gambling	☐ Broken the law or considered breaking the law to finance gambling
☐ Other:	☐ Gambled to get money to meet financial obligations

Risk Taking/Impulsive Behavior (current/past) – select all that apply:

☐ Unprotected sex	☐ Shoplifting	☐ Reckless driving
☐ Gang Involvement	☐ Drug Dealing	☐ Carrying/using weapon
☐ Other:		

BIOPSYCHOSOCIAL ASSESSMENT

Mental Status Exam

Category	Selections			
GENERAL OBSERVATIONS				
Appearance	☐ Well groomed	☐ Unkempt	☐ Disheveled	☐ Malodorous
Build	☐ Average	☐ Thin	☐ Overweight	☐ Obese
Demeanor	☐ Cooperative	☐ Hostile	☐ Guarded	☐ Withdrawn
	☐ Preoccupied	☐ Demanding		☐ Seductive
Eye Contact	☐ Average	☐ Decreased		☐ Increased
Activity	☐ Average	☐ Decreased		☐ Increased
Speech	☐ Clear	☐ Slurred	☐ Rapid	☐ Slow
	☐ Pressured	☐ Soft	☐ Loud	☐ Monotone
	Describe:			
THOUGHT CONTENT				
Delusions	☐ None Reported	☐ Grandiose	☐ Persecutory	☐ Somatic
	☐ Bizarre	☐ Nihilist		☐ Religious
	Describe:			
Other	☐ None Reported	☐ Poverty of Content	☐ Obsessions	☐ Compulsions
	☐ Phobias	☐ Guilt	☐ Anhedonia	☐ Thought Insertion
	☐ Ideas of Reference		☐ Thought Broadcasting	
	Describe:			
Self Abuse	☐ None Reported		☐ Self Mutilization	
	☐ Suicidal (assess lethality if present)		☐ Intent	☐ Plan
Aggressive	☐ None Reported	☐ Aggressive (assess lethality of present)		
	☐ Intent		☐ Plan	
PERCEPTION				
Hallucinations	☐ None Reported	☐ Auditory		☐ Visual
	☐ Olfactory	☐ Gustatory		☐ Tactile
	Describe:			
Other	☐ None Reported	☐ Illusions	☐ Depersonalization	☐ Derealization
THOUGHT PROCESS				
☐ Logical	☐ Goal Oriented		☐ Circumstantial	☐ Tangential
☐ Loose	☐ Rapid Thoughts		☐ Incoherent	☐ Concrete
☐ Blocked	☐ Flight of Ideas		☐ Perserverative	☐ Derailment
Describe:				
MOOD				
☐ Euthymic		☐ Depressed		☐ Anxious
☐ Angry		☐ Euphoric		☐ Irritable
AFFECT				
☐ Flat	☐ Inappropriate	☐ Labile		☐ Blunted
☐ Congruent with Mood	☐ Full		☐ Constricted	
BEHAVIOR				
☐ No behavior issues		☐ Assaultive		☐ Resistant
☐ Aggressive		☐ Agitated		☐ Hyperactive
☐ Restless		☐ Sleepy		☐ Intrusive
MOVEMENT				
☐ Akathisia	☐ Dystonia		☐ Tardive Dyskinesia	☐ Tics
Describe:				
COGNITION				
Impairment of:	☐ None Reported		☐ Orientation	☐ Memory
	☐ Attention/Concentration		☐ Ability to Abstract	
	Describe:			
Intelligence Estimate	☐ Mental Retardation	☐ Borderline	☐ Average	☐ Above Average
IMPULSE CONTROL		☐ Good	☐ Poor	☐ Absent
INSIGHT		☐ Good	☐ Poor	☐ Absent
JUDGMENT		☐ Good	☐ Poor	☐ Absent

BIOPSYCHOSOCIAL ASSESSMENT

RISK ASSESSMENT

Risk to Self	☐ Low	☐ Medium	☐ High	☐ Chronic
Risk to Others	☐ Low	☐ Medium	☐ High	☐ Chronic

Serious current risk of any of the following: (Immediate response needed)

Abuse or Family Violence ☐ Yes ☐ No	Abuse or Family Violence ☐ Yes ☐ No	
Psychotic or Severely Psychologically Disabled	☐ Yes ☐ No	
Is there a handgun in the home? ☐ Yes ☐ No	Any other weapons? ☐ Yes ☐ No	
Plan:		
Safety Plan Reviewed ☐ Yes ☐ No		

Diagnoses and Interpretive Summary

Biopsychosocial formulation

DSM IV-TR Provisional Diagnoses

Axis I	
Axis II	
Axis III	
Axis IV	
Axis V	

Treatment Acceptance/Resistance

Client accepts problem?   ☐ No   ☐ Yes   Comment:
Client recognizes need for treatment? ☐ No   ☐ Yes   Comment:
Client minimizes or blames others?   ☐ No   ☐ Yes   Comment:
External motivation is primary?   ☐ No   ☐ Yes   Comment:

Strengths/Resources   (enter score if present)   **1 = Adequate, 2 = Above Average, 3 = Exceptional**

Family Support	Social Support Systems	Relationship Stability
Intellectual/Cognitive Skills	Coping Skills & Resiliency	Parenting Skills
Socio-Economic Stability	Communication Skills	Insight & Sensitivity
Maturity & Judgment Skills	Motivation for Help	Other:

Comments:

Describe appropriateness & level of need for the family's participation:

Revised 5/3/06

BIOPSYCHOSOCIAL ASSESSMENT

Preliminary Treatment Plan & Referrals

Preliminary Biopsychosocial Treatment Plan
Biological:
Psychological:
Social/Environmental:

Referrals			
☐ Psychiatrist	☐ Psychologist	☐ Medical Provider	☐ Spiritual Counselor
☐ Benefits Coordinator	☐ Nutritionist	☐ Rehabilitation	☐ Vocational Counselor
☐ Social Worker	☐ Community Agency:		☐ Other:

Physical Fitness (Optional)

Physical Activity (please select one of the following based on activity level for the past month):

☐ Avoids walking or exertion, e.g. always uses elevator, drives whenever possible instead of walking.

☐ Walks for pleasure, routinely uses stairs, occasionally exercises sufficiently to cause heavy breathing or perspiration.

Participates regularly in recreation or work requiring **modest physical activity** such as golf, horseback riding, calisthenics, gymnastics, table tennis, bowling, weight lifting, and yard work.
☐ 10-60 minutes per week
☐ More than one hour per week

Participates regularly in **heavy physical exercise,** such as running, jogging, swimming, cycling, rowing, skipping rope, running in place or engaging in vigorous aerobic activity such as tennis, basketball or handball.
☐ Runs less than a mile a week or engages in other exercise for less than 30 minutes per week
☐ Runs 1-5 miles per week or engages in other exercise for 30-60 minutes per week
☐ Runs 5-10 miles per week or engages in other exercise for 1-3 hours per week
☐ Runs more than 10 miles per week or engages in other exercise for more than 3 hours per week

Revised 5/3/06

OTHER BOOKS BY THE AUTHOR

CHALLENGING MY FAITH 1 & 2

By Very Rev Joyzy Pius Egunjobi

CHALLENGING MY FAITH (Questions People Ask) is a book that treats questions of faith in a way to help those who are confused about their Christian and Catholic Faith. It consists offour volumes: Volume 1: The Holy Trinity (Father, Son and Holy Spirit) Volume 2: The Church, Mary and Joseph and The Saints.

CONQUERING DEATH

By Very Rev Fr Joyzy Pius Egunjobi

When Jesus promise LIFE, it is life as opposite of Second Death – thus He says in John 6:57, "Just as the living Father sent me and I live because of the Father, so the one who feeds on me will live because of me". Then the end will come, when Jesus hands over the kingdom to God the Father after he has destroyed all dominion, authority and power. For Jesus must reign until he has put all his enemies under his feet. The last enemy to be destroyed is death. - 1Cor 15:24-26. Then I will die no more but live to praise my God forever and ever.

IN LOVE WITH A GHOST
By Joyzy Pius Egunjobi
Bidemi Adewale had been very faithful and sincere to his priestly vocation. At his first glimpse of Mrs. Angela Ukanwa, he became another person. He started doing what he had never done before: He knew what it is to make love. On his return to the Seminary, he could not stay - "I have been unfaithful...I have failed the Lord", he said. He left the Seminary. Two weeks to his wedding, he discovered that there was a Mystery surrounding the being of his wife-to-be; "My girl is a Ghost...". Would he continue with the wedding?. If he did not, how did he go about it? If he did, how did he cope?

NOT BECAUSE OF ME

By Joyzy Pius Egunjobi

Bidemi and Angela had only two children apart from Jean. His son, Bidemi (Jr.) went into priesthood and was accused of attempted rape to his father's friend, Bola Bassey. Jean had an unlawful affair with an Italian lady, Margaretto. Jean had mysterious twins: Peter and Paul. Peter, a priest and Paul in bad company of Andrew. Peter saw it as his mission to bring both Paul and Andrew into the light of Christ. Andrew's success hinged on his plan to distribute pictures of Fr. Peter and a lady in compromising circumstances. He was to do this on a Monday. That Sunday, Andrew was murdered. Patrick, a mission boy and a nephew to Andrew, claimed strongly that Fr. Peter is a suspect. The court case was a confusing one. Fr. Peter, a popular spiritual father, was found guilty of murder and thus sentenced to death. What did Paul mean when he shouted, "No....Not him....NOT BECAUSE OF ME, as his twin brother was being led away?

THE TRUTH OF LIFE

By Joyzy Pius Egunjobi

The thoughts as presented in this collection are not mere English grammar or constructions of sentences but thoughts with deep meanings, thoughts, which can be fully fathomed by great minds. Therefore, they are best comprehended with easy but thinking reading.

THE PASTORAL CARE OF THE PHYSICALLY CHALLENGED

By Rev Joyzy Pius Egunjobi

It is with great interest that I read this book written by Rev Fr Joyzy Pius Egunjobi of the Catholic Diocese of Oyo in Nigeria. This is because of my passion to enrich my own knowledge in handling issues relating to persons with disabilities. What this book succeeded in pointing out to me as I read from page to page is the undesirable gap that human beings have created between the living word of god as reflected in the different chapters where many scriptural passages were copiously quoted and the way majority of people live and deal with neighbors, even after knowing that we are all created in the image and likeness of God.

THINGS TO CONSIDER BEFORE STEPPING INTO MARRIAGE

By Rev Joyzy Pius Egunjobi

At this time when many of our youths are becoming more responsive to marriage, there could be no better material on marriage than this challenging and thought-provoking publication THINGS TO CONSIDER BEFORE STEPPING INTO MARRIAGE. Although, the title of this book appears commonplace and simplistic, a thorough, careful and meditative reading of the whole content will definitely convince the reader that there is far more to the publication than the title indicates. Through theological, philosophical, socio-cultural and scientific approach, the author, Rev Joyzy Pius Egunjobi treats the topic down-to-earth.

THE SECRETS OF A HAPPY MARRIAGE

By Rev Joyzy Pius Egunjobi

At a time like this when Marriage which is supposed to be a Happy encounter and lovely experience is now for many a situation of hatred, tears, regret, and irresponsibility Rev Joyzy Pius Egunjobi makes us see that all the ugly experiences in marriage are due to the fact that we have not unveiled the SECRETS OF MARRIAGE

THE POLICE GOD

By Rev Joyzy Pius Egunjobi

This book will be useful to the philosopher who wants to increase his knowledge about God's relationship with man in the area of law, order and security. It will be useful to the policeman who is being asked to bring in God into the daily performance of his duties. From members of the public, the book solicits for more understanding for some of the frailties and human mistakes committed by the police in the course of his duties.

SUBSTANCE ABUSE AND THE YOUTHS: CHRISTIAN PERSPECTIVES

By Joyzy Pius Egunjobi

Substance Abuse and the Youths: Christian Perspectives does provide you with the knowledge and tools one will need to begin the battle with abusive substances. Not only will it share with you its origin, etiology, and effects, it will also show the damages that will occur to you physically, spiritually and emotionally. The information contained in the book was enlightening and easy to read and understand. As a Christian this book can and will put substance abuse in perspective with our belief. I recommend that Christians and Non- Christians alike read this book especially if they have children or know anyone who uses or abuses substances.

GENETICS AND ADDICTION

By Dr Joyzy Pius Egunjobi

Undoubtedly, from genetics and neurological outlook, addiction is a heritable disease that can be transmitted from parents to children and of course from one generation to the other. According to Thich Nhat Hanh "If you look deeply into the palm of your hand, you will see your parents and all generations of your ancestors. All of them are alive in this moment. Each is present in your body. You are the continuation of each of these people". Yet, the environmental factors of addiction cannot be overlooked as they influence and interact with genetics - in other words, In the success of nature we are nurtured, and in the success of nurture we become who we are - This is what this book expresses in the understanding and answering the question: WHY DO SOME PEOPLE GET ADDICTED AND SOME OTHERS NOT?

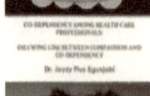

CO-DEPENDENCY AMONG HEALTH CARE PROFESSIONALS: DRAWING LINE BETWEEN COMPASSION AND CO-DEPENDENCY

By Dr. Joyzy Pius Egunjobi

Counseling is helping profession, counselors are caring and compassionate, and counselors like any other human beings are susceptible to codependency. Yet, counselors have all that it takes to show compassion without enabling or slipping into codependency as long as they are aware of their own issues, not trying to be controlling and not trying to fix others' problems or taking responsibility for the problems of others that are not theirs. Dr Joyzy Pius Egunjobi has, through this book, systematically guide those in health profession on the need to be aware of the very thin line between showing compassion and becoming codependent. By becoming aware of the signs and symptoms of codependency, helpers will be able to take precautions and learn how to take care of their own physical and mental health. It is a call to health professionals: "Physician, heal yourself"

MARIJUANA LEGALIZATION VERSUS MARIJUANA ADDICTION: THE POLITICS
By Rev. Dr. Joyzy Pius Egunjobi

The question Can legalization of marijuana perpetuate marijuana addiction? has in recent times generated so much debates with conflicting scientific research findings. No doubt, the helm plant called cannabis or marijuana has been and is very useful to man in different capacities ranging from nutritional or as food consumption; to its medical use to increase appetite and reduces nausea, decrease pain, inflammation, muscle control problems, controlling epileptic seizures, and possibly even treating mental illness and addictions.
Notwithstanding, the use of marijuana has some negative effects on physical as well as mental health. Should it be legalized considering the medical use and the financial gain the nation can derive from it; or should the nation still regard it as Schedule 1 Drug, because of its destructive effects. This is What Rev Dr. Joyzy Pius Egunjobi seeks to resolve in this well researched work.

21 DAYS NOVENA OF MY MEDITATION OF THE SCRIPTURAL ROSARY
By Rev Fr Joyzy Pius Egunjobi &

Rev Fr. Julius Olaitan

21 DAYS NOVENA OF MY MEDITATION OF THE SCRIPTURAL ROSARY is a novena book that guides Catholics and non-Catholics in obtaining an increase of

faith, hope and love; leading to an intense love forJesus Christ through the powerful intercession of Mary, Mother of God. With daily prayers that include biblical passages for meditation and spiritual growth, it helps one towards an attainment of personal petitions The writers of this book are seasoned pastors who have read wider. They have arrived at this point to bring their children in the faith to better understanding of their faith. The meditations before the recitations are simply superb. They bring the person(s) so close to the mind of the Church since earlier times. This book contains some other prayer points which the users can use to know more about the catholic faith and beliefs.

BE BLESSED: Reflections and Christian Teachings

By Rev. Joyzy Pius Egunjobi

"Be Blessed" is a product of daily online ministry where numerous people are blessed every morning/afternoon/evening, depending on the location on the globe, with Scriptural passages, personal reflections, answers to questions, and Christian teachings. When we experience hurt, anger, prejudice, pain, suffering, separation, divorce, poverty, loneliness, spiritual dryness to mention a few; we often feel sad, depressed, hopeless and overwhelmed. Fr Joyzy Pius Egunjobi has through this work, made it possible to experience the value of Psycho – Spiritual Therapy and the power of God's Healing Love. Remember to "sanctify the Lord God in your hearts: and be ready always to give an answer to every man that asked you a reason of the hope that is in you with meekness and fear..." (1Pet 3:15) Be Blessed

BE BLESSED 2: Questions of Faith and Sacraments

"Be Blessed 2: Questions of Faith and Sacraments" is a product of continued daily online ministry of answering faith questions with more than 2.5 million daily recipients. When we experience hurt, anger, prejudice, pain, suffering, separation, divorce, poverty, loneliness, spiritual dryness to mention a few; we often feel sad, depressed, hopeless and overwhelmed. Fr Joyzy Pius Egunjobi has through this work, made it possible to experience the value of Psycho –

Spiritual Therapy and the power of God's Healing Love. It is a must read book to sanctify each day.

BOOKS FROM THE SAME AUTHOR

1. In Love With a Ghost
2. Not Because of Me
3. The Secrets of a Happy Marriage
4. Things to Consider Before Stepping Into Marriage
5. The Police God
6. The Truth of Life: Quotable Quotes
7. The Truth of Life – Special Edition
8. Substance Abuse And The Youth: Christian Perspective
9. The Pastoral Care of the Physically Challenged
10. The Church I Belong To
11. My Meditation of The Lord's Prayer
12. Challenging my Faith Vol 1
13. Challenging my Faith Vol 2
14. Challenging my Faith Vol 1&2 Combined Edition
15. Conquering Death
16. Be Blessed: Reflections & Christian Teachings
17. Be Blessed 2: Questions of Faith and Sacraments
18. 21 Days Novena of my Meditation of the Scriptural Rosary
19. Genetics and Addiction
20. Co-Dependency Among Health Care Professionals: Drawing Line Between Compassion And Co-Dependency
21. Marijuana Legalization and Legalization Addiction: The Politic.

Available in *Amazon, Ebay, Itunes, Playstore, and many online Bookstores in paperback or ebook format*

Also available at www.lulu.com/joyzypbooks

ABOUT THE AUTHOR

 ***Rev. Dr. Joyzy Pius Egunjobi*** is a priest of the Catholic Diocese of Oyo, Nigeria and a seasoned Psycho-Spiritual Therapists, a Trainer with Psycho-Spiritual Institute, Kenya, and an Addiction Specialist.

He is a Philosophical, Theological and Psycho-Spiritual Poet and Author of more over 20 books. His passion for educating the youths and adults greatly inspired his writing talents.

He loves writing, traveling, music, and educating people on the reality of life.